Congmom's Cake Diary

Cake Design Recipe

Jung Hayeon

Jung Hayeon

Majored in architecture and taught math, and now delivering whipped cream cake design skills. Experienced various environments and variables through domestic and overseas lectures, and now runs the 'Congmom Cake' Hanok Studio in Hyehwa-dong and teaches design skill classes for whipped cream cakes. The books written includes 『Congmom's Cake Diary 2: Cake Design Recipe』 and 『Congmom's Happy Baking Diary』.

- **Instagram** @congmom_cake
- **Blog** http://blog.naver.com/wls7912

Cake Design Recipe

: Congmom's Cake Diary

First edition published	December 21, 2020
Second edition published	July 29, 2022
Author	Jung Hayeon
Translated by	Kim Eunice
Publisher	Han Joonhee
Published by	iCox, Inc.
Plan & Edit	Bak Yunseon, Jang Areum
Design	Chang Jiyoon, Lee Jisun
Photograph	Kim Namhun
Sales/Marketing	Kim Namkwon, Cho Yonghoon, Moon Seongbin
Management support	Son Okhee, Kim Hyoseon
Address	Jomaruro 385beongil 122, Sambo Techno Tower, Unit 2002, Bucheon-si, Kyeonggi-do, Republic of Korea
Website	http://www.icoxpublish.com
Instagram	@thetable_book
E-mail	thetable_book@naver.com
Phone	82-32-674-5685
Registration date	July 9, 2015
Registration number	386-251002015000034
ISBN	979-11-6426-156-7 (13590)

Cake Design Recipe

_Prologue

From a student of architecture who used to expand her imagination making models with a 30° cutter knife and rising paper to a mathematics teacher in my early 30s, I have spent each day full of dreams and energy. Even though I have entered the path of full-time housewife after getting married and giving birth, I was a mom full of excitement, who would write down the schedules every day and spend the day in full even if I was hanging around.

It was when my baby was 16-month-old. I was able to send my baby to a daycare center earlier than expected; the full-time housewife finally got the free time. The time given in a day was about five hours. I didn't want to spend the time in vain, so I enrolled in the 'Bread & Cookie 101'.

That's how it began at a little late age of 34. I enjoyed making cookies and macarons. But when I started to enjoy making especially cakes, I realized that I needed more skills to expose my thoughts and express them freely. Even in architecture, it was possible to convey the ideas only when described in pictures, programs, and models.

As I was studying the cake designer course, I was confident that my area of interest was 'fresh cream' once more. My interest in decorations using piping tips grew more profound, and I started to teach my first cake design class in March of 2017.

There were many classes teaching tasty recipes, but there were not enough classes teaching how to express your thoughts and make beautiful cakes. I was so curious about it, and I thought, 'Why don't I try it!'. The class, which started as a one-day class, gradually became a regular class of 3-week course, and in the winter of that year, a new standard class was created called Event Cake Course.

A year has passed, and I had a lot of opportunities to try different challenges in 2018. I finally opened a studio and taught abroad in China, Taiwan, etc. It was amazing that many countries showed so much interest in whipped fresh cream cake designs; hence, the sense of responsibility increased even more.

When I started to think about making a book that contains Congmom's skills and tips and photos of the working process, several publishers began to contact me. Working with the team manager, Bak Yunseon, who read my stories from the beginning of Congmom until now, and checking thoroughly, continued from spring to winter.

Congmom's Cake Design is a class focused on skills. What is being conveyed is essential, but I think how you say is more important. I refined this class by studying every day, speaking out loud to practice, and practicing the method of execution.

Through this book, I hope that those entering into cake design for the first time will gain interest and approach easily, and to those in the field can train their ability to observe and that I can provide tips one by one.

Now that the year is coming to an end and getting ready for the brand new year, my heart wiggles while making plans for next year. There are so many things I want to do and learn. I want to accomplish each one by sharing my thoughts with the people who love cake as we do now.

Thank you sincerely to my husband- Siwon's dad, who supports my dream; my son, my endorphin Siwon; and my friend Joy, who has always been by my side since the beginning of my first lecture. It took a lot of effort and sincerity to make this one book. Most of all, I am grateful to the members of the publisher TheTable, who have given me the opportunity to look back on what I do, to sort out and plan for the days to come.

December 2020, Jung Hayeon

How to use this book

LESSON 01.

Basic tools & Ingredients

This chapter describes the tools and materials mainly used in this book. The design of the whipped cream cake is completed in a wide variety of designs depending on the tools and ingredients used. Do understand the characteristics of each tool and ingredient and apply them in various designs.

LESSON 02.

Making Whipped Cream for Decoration

This chapter explains the most important ingredient, whipped cream. The theory and process of making whipped cream suitable for its use and the whipped cream recipe are explained easily. The density of whipped cream is explained step by step to finish whipping according to the desired density.

LESSON 03.

Cutting Sponge Cakes and Filling with Whipped Cream

This chapter describes how to cut sponge cake according to the shape of the cake and how to fill with cream depending on the presence of fruits. It is a process that needs to be prepared well for a full-fledged cake design.

LESSON 04.
Icing

The design class starts. Depending on the shape of the cake, the method of icing depends on the tool used. In this chapter, the icing techniques of round, dome, chiffon, square, and heart-shaped cakes are explained in detail with images of the process. For those who have difficulty with icing, an explanation of how to ice easily using a scraper is included.

The appropriate density and level of difficulty for each icing are indicated.

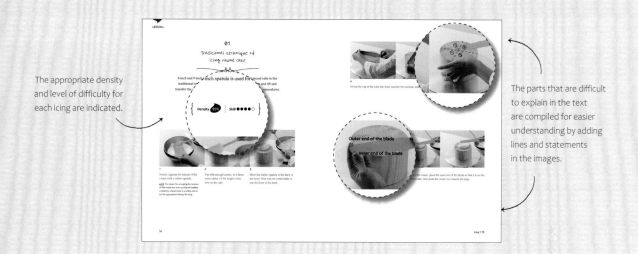

The parts that are difficult to explain in the text are compiled for easier understanding by adding lines and statements in the images.

LESSON 05.
How to Restore and Finish Icing

Icing can sometimes lead to partially incomplete finishing. Tips are offered on how to complete with only partial recovery, without having to start all over again. It explains how to safely transfer the finished cake from the turntable to the cake board and complete it neatly.

Shows both the wrong and the right way to help to avoid making mistakes twice.

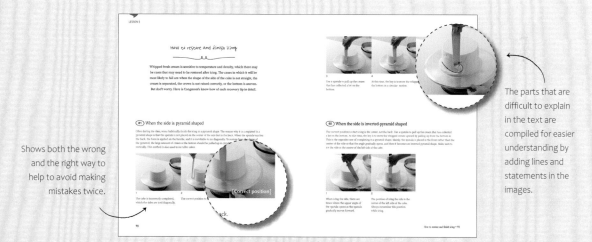

The parts that are difficult to explain in the text are compiled for easier understanding by adding lines and statements in the images.

LESSON 06.

Decorating with Piping Nozzles

There are endless decoration techniques using piping nozzles. Depending on the type of piping nozzles and techniques used, it can be finished in a wide variety of shapes and applied to cake design. In this chapter, you can learn about a variety of piping nozzles and techniques, ranging from the most frequently used closed star nozzles, round nozzles, French star nozzles to leaf nozzles, petal nozzles, and drop flower nozzles.

Depending on the piping nozzle and technique, the density, angle, and starting position will vary. Before beginning to decorate with piping nozzles, key points are prepared to recognize at once.

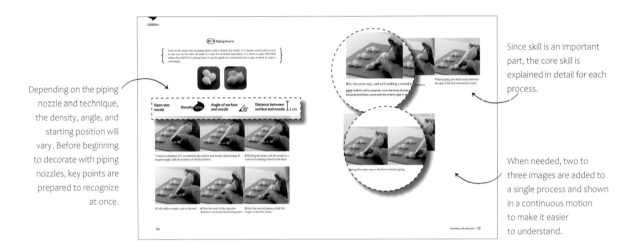

Since skill is an important part, the core skill is explained in detail for each process.

When needed, two to three images are added to a single process and shown in a continuous motion to make it easier to understand.

LESSON 07.

Decoration Techniques Using Other Tools

In addition to the piping nozzles, a variety of cake designs can be created using a spatula, measuring spoon, fork, brush, etc. Learn and apply various techniques- from making cute peaches to shaping petals used to cover the top of a chiffon cake.

Tools used and the level of difficulty is indicated.

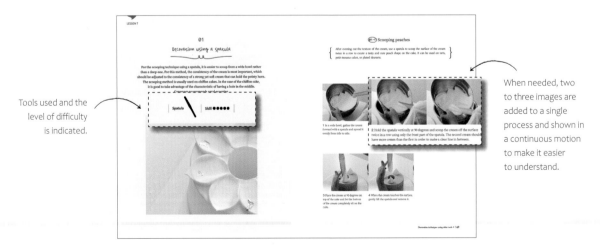

When needed, two to three images are added to a single process and shown in a continuous motion to make it easier to understand.

LESSON 08.

Congmom's Cake Diary

Thirty-seven kinds of Congmom's whipped fresh cream cake designs loved by many people are introduced. Based on the various techniques learned in previous chapters, it is completed by combining different ingredients and decorations suitable for the design. Congmom's Cake Design Class know-how has been transferred as is.

The tools and techniques used in each cake design, starting angle of piping, piping cream density, and level of difficulty are indicated.

This is the page sign icon. If a detailed explanation of this process is needed, go back to the page indicated on the icon and check the detailed theory.

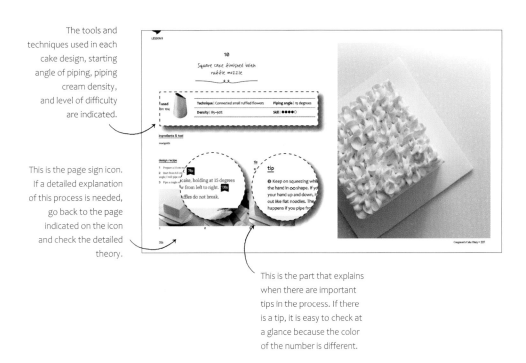

This is the part that explains when there are important tips in the process. If there is a tip, it is easy to check at a glance because the color of the number is different.

Reminder

The 'names of the techniques' used in this book are actually utilized in Congmom's Cake Design Class. Please note that the author Jung Hayeon has given names to the techniques without a formal name, and there are some techniques she created herself- these names are given to help understand better.

Cake design is

It's not easy to define 'cake design' in one word. Still, it means that visual beauty and taste stimulation can be drawn simultaneously by integrating various elements such as shape, size & composition, color & decoration. These days, cakes with more unique designs are pouring out, besides the usual shapes such as round, dome, and square.

While mold-based cakes like mousse cakes may be freer from formative expressions, whipped cream cakes can also be expressed in various forms. In this book, icing techniques of different shapes and various decorating methods will be covered in detail using different tools and tips.

Contents

Contents

LESSON

07

Decoration Techniques Using Other Tools

Contents

LESSON

PLUS

Sponge Cake Recipe

LESSON 01

Basic Tools &
Ingredients

This chapter will introduce the tools and ingredients necessary for the design of the whipped fresh cream cake. As it can be completed with various designs depending on the tools and ingredients, it is essential to understand the characteristics, strengths, and weaknesses of the tools and ingredients to produce high-quality designs.

⓪1 Basic tools

1. Turntable

Expert-size (31 cm diameter, 10 cm height) cast iron base turntables are best to use. The base is covered with rubber, which prevents slipping, and it is made of cast iron, which provides weight for stability. The three lines on top help to set the center according to the size of the cake. A regular turntable is easy to wobble when used for an extended time. Plastic turntables are more suitable for flower cakes or fondant cakes than icing whipped cream cakes.

2. Spatula

8-inch and 9-inch spatula are mainly used for 15 cm and 18 cm cakes. Spatula produced in Japan is light and thin, which is best for icing light whipped cream cake. As for the spatula made in Korea, although the width of the blade is narrow because the tip of the blade is round, it works well on the stainless-steel turntable. Generally, an 8-inch spatula is used for icing, and a 9-inch spatula is used to clean up the bottom of the cake and to transfer. It is recommended to replace the spatula often because the blade can bend after using it for a long time due to the weakness of its thin and narrow base.

3. Wood mock-up

It is a model to practice icing. The tall and round, chiffon, square, and heart-shaped models introduced in this book are all made in a woodworking shop in a diameter of 15-centimeters. Off-the-shelf products can be purchased at Bangsan market or online shopping malls. (May be limited to Korea.)

4. Stainless-steel tray

It is good to place the mainly used items while working, such as small tools, fruits, or flowers for decoration.

5. Rubber spatula

It is used when organizing the texture of the cream evenly and when scooping to transfer the whipped fresh cream. A spatula with a thin blade is suitable for handling whipped cream.

6. Stainless-steel bowl

It is used to whip fresh cream. It is better to use a wide bowl that is easy to remove the whipped cream with a spatula or rubber spatula than a deep U-shaped bowl.

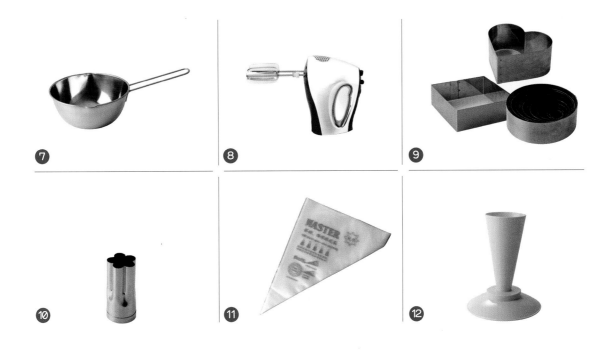

7. Double boiler pot

When using just a little bit of ingredients such as ganache used for the melting-cake and drip-cake, it is convenient to use a small saucepan with a handle.

8. Hand mixer

It is used to give volume to fresh cream. It has the advantage of being able to work easily in a short time. A model in which speed can be adjusted in increments is convenient.

9. Mousse ring

The mousse ring used to cut shaped sheets is usually 15 cm in diameter. For stacking type of cakes, such as a tree-shaped cake, it is convenient to use rings of different sizes so that the diameter of the sheets can be reduced by 1.5~2 cm.

10. Flower shaped cutter

It is used to cut fruits in flower shapes to decorate. Try using other desired shapes of cutters such as heart or star.

11. Piping bags

It is used to contain whipped cream and to pipe. A 14-inch piping bag is suitable for whipped dairy cream. If the piping bag is too big, the cream can separate easily during putting it in and organizing it. Put the nozzle in the piping bag, leave 2/3 of the length of the nozzle, and cut the front part to use. Disposable piping bags are recommended to use. It is significantly better to use the piping bag without a seam line from top to bottom because the seam line won't touch the cake or the piped cream when decorating.

12. Piping bag holder

It is convenient for holding piping bags filled with whipped cream or chocolate. It can fit 14-inch and 16-inch piping bags and also suitable to put cream inside. It is easy to store and clean when the bottom plate is detachable.

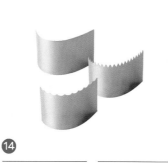

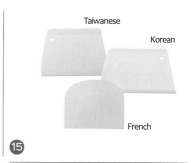

Taiwanese

Korean

French

⑬

⑭

⑮

⑯

⑰

13. Piping nozzles

These are used to pipe cream into various shapes on the iced cake. Depending on the shape of the nozzle, it can be decorated with different shapes such as teardrops, hearts, flowers, etc. For this book, domestic nozzles that are easy to purchase were used.

14. Dome cake scraper

Three different dome cake scrapers are usually used to make dome cakes. Since ready-made scrapers have narrow and wide curved angles, it is recommended to open the angle of the scraper by hand according to the height of the cake.

15. Scrapers

It is used to ice the side of the cake along with the spatula. There are scrapers made in Korea, Taiwan, France, and Japan, and knowing the characteristics of scrapers is convenient to use accordingly. In the case of a Korean scraper, the bottom has to be leveled and wears out quickly. Taiwanese scraper is easy to use, so it is recommended for beginners. French and Japanese scrapers have sharp edges, so it is good to use for finishing after icing.

16. Cake combs

It is a scraper that can highlight the side of the cake. There are many kinds, from decorative design to simple and plain patterns.

17. Cake pan

These are used for baking genoise and chiffon. For cake design purposes, cakes are often baked in tall cake pans than low pans. Because Korean chiffon pans are low in height, taller Japanese pans are mainly used. It is convenient to use a round pan with a removable bottom.

(18)

(19)

18. Fruit-baller (Fruit scoop)

Fruits that the flesh can be shaped, such as mango, melon, and watermelon, can be cut with cookie cutters or dig out with a fruit scoop and used as decoration. It is used to carve out fruits in cute round shapes, and 22 mm in diameter is best to decorate a 15 cm cake. Try using different sizes of fruit scoops to make it more interesting.

19. Decoration brush

When placing gold or silver leaf on top of the cream, use a small brush with stiff hair. The brush is more convenient to use than tweezers when removing the gold leaf. For the brush used to decorate the side of the iced cake, it is convenient to use a wide brush that is light, with a short handle that is easy to hold.

(02) Essential ingredients- Fresh cream

1. Fresh dairy cream

It uses only natural fat and has a unique soft flavor, but the milk fat structure itself is weak, making it easy to get rough. The fresh dairy creams used in this book are 38% fat Seoul Milk fresh cream, Maeil Milk fresh cream, and 45% fat Pasteur fresh cream. Most of the fresh domestic cream is set at 35% milk fat content. The milk fat content has gradually increased in line with the recent trend of more people liking the rich flavor. The higher the milk fat content, the taste of milk is deep and rich, and has a creamy mouthfeel. In the case of fresh dairy cream, it is recommended to use it after comparing per brand since there is a difference in the progress of glossiness and when it starts to get rough from the 80% whipped stage.

2. Non-dairy cream

It is a cream made mainly of vegetable fats such as corn oil, cottonseed oil, soybean oil, and palm oil. It is more durable and cheaper than fresh dairy cream but lacks rich flavor. However, it is suitable for practicing because it is easier to ice and pipe with nozzles. Maeil Whippee was used for this book. There are other products such as Ever-Whip and Gold Label, and it can be mixed with shortening and powdered sugar.

(1)

(2)

⑳③ Decoration ingredients

1. Fresh fruits

It is the ingredient that goes best with a whipped cream cake. The most commonly used fruits are red-toned fruits that have definite contrast with white whipped cream, which includes strawberries, raspberries, grapefruits, peaches, watermelon, cherries, pomegranate, red currant, mini apples, and figs. Mainly used yellow-toned fruits are mango, gold kiwi, lemon, banana, and green-toned fruits consist of kiwi, melon, and green grapes. Dark-toned fruits are also used, such as blueberries and black currents.

2. Edible flowers

There are fresh flowers such as pansy, nasturtium, edible rose, chrysanthemum, and dried edible flowers such as cornflower, calendula (marigold) and cherry blossoms. These days, a variety of edible flowers are easily found at grocery stores or online shopping malls.

3. Herbs

These include thyme, rosemary, apple mint, lavender, baby leaves, basil, dill, etc. Herbs with dark leaves or leaves that are spaced too tight are not suitable for decoration. As for thyme, the thin and naturally stretched stem looks more beautiful when placed on the cake.

4. Gold leaf sheets and silver leaf sheets

It is used to give an accent on fruits, flower petals, or herbs. There are gold leaf sheets and silver leaf sheets that can be cut and used as desired, and there are gold and silver flakes that are cut into small pieces and contained in a case or bottle. Gold leaf sheets and silver leaf sheets are good to use for special fancy cakes.

5. Roasted nuts and gold powder

Roasted nuts rolled in the gold powder can be used as excellent point decoration. It goes exceptionally well with a winter-themed cake.

6. Chocolate decoration

Bar chocolates that can be broken or chocolate stamped with cookie cutters are good to use as point decorations on a cake.

7. Croquant

Chocolate and textured ingredients can be mixed and used as cake decoration. Not only for the visual effect, but also tastes good, so it is versatile to use widely.

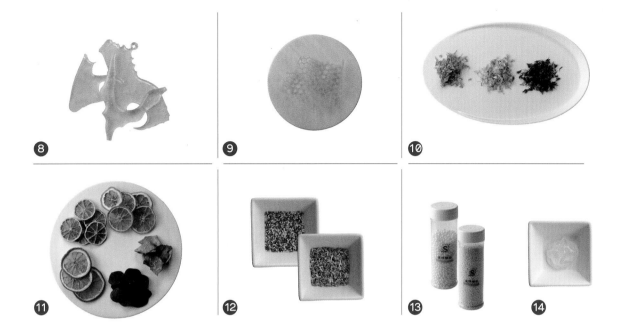

8. Isomalt chip

Isomalt is a type of sugar that can be baked in an oven or heated in a pot and make into desired shapes. Isomalt chip is transparent, and has a fancy shape and reflects light according to the viewing angle, which can create a sophisticated design even with one piece.

9. Honeycomb decoration

It is a simple decoration made with cake flour, egg whites, and butter, which goes well, which goes well as a decoration for pumpkin cake. It can also be made with chocolate.

10. Dried petals

Small dried petals are good to use to accentuate a tidy cake without being excessive. Marigolds and cornflowers are often used. It can give a good effect by itself without having to use other ingredients.

11. Dried chips

Sweet pumpkin, purple sweet potato, orange, lemon, apple, and others are available, and they are used mainly on chiffon cakes. Especially the redlove chips, which are dried apple with red flesh, give off deep red color that creates a gorgeous effect.

12. Sprinkles

Sprinkles, often used in confectionery, such as cookies, muffins, and donuts, can be finely ground and mixed with whipped cream to give an interesting and colorful effect. When purchasing sprinkles for the icing, avoid products that bleed and melt when mixed with whipped cream.

13. Arazan sugar pearls

Arazan pearls are mainly used for Christmas tree cakes because it goes well with a winter theme. It also goes well with lovely-themed cakes.

14. Decogel (mirror glaze)

It forms a thin and transparent layer on the surface of the cake, such as mousse cakes, to make them look moist. When it is applied to the decorated petals, it gives the effect of a water drop. It can be piped with a piping bag or pipe it a little bit at a time with a cornet made with an acetate film.

LESSON 02

Making Whipped Cream for Decoration

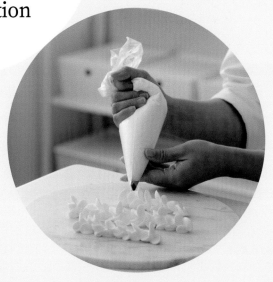

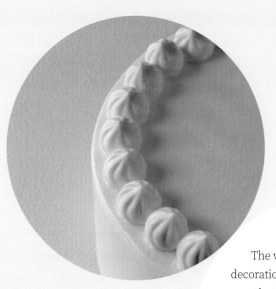

The whipped cream used for decoration can be divided into insert cream that fills between the sheets, icing cream covering the surface of the cake, and decoration cream that decorates the iced cream. Let's learn about the types and characteristics of whipped cream and find out the appropriate density for each whipped cream.

(01) Preparing fresh cream

The cream used for a whipped cream cake can be divided into insert cream that fills
between the sheets, icing cream covering the surface of the cake, and decoration cream
that decorates the iced cream. Please remember that each cream has
a different whipping consistency.

1

In a big bowl, add 160 g insert
cream, 180 g icing cream (total
340 g of cream), and 34 g sugar.
Place the bowl over an ice-bath.

point It is based on a 15 cm cake where
fruits will be filled between the sheets,
and using sugar of 10% of fresh cream.
If larger (by height) fruits are used,
increase the amount of fresh cream and
sugar.

2

Set the speed of the hand mixer to
low (1st speed) to dissolve sugar and
start whipping.

3

When the bubbles start to appear all
over, increase the speed to 2nd or 3rd
speed to create volume.

4

Whip up to 70% and lower to 1st speed to even out the texture of the cream.

point When the cream is scooped and dropped with a spatula, a line appears around the dropped cream and disappears immediately- this is cream whipped to 70%.

5

Prepare to scale 180 g cream for icing in a separate bowl. Wipe the water underneath the bowl to prevent water from getting in.

6

Place a bowl on a scale and tare to 0 (zero). When pouring the cream from the bowl, be careful not to touch the cream with the spatula, and control the amount by tapping the edge of the bowl with the longer blade of the spatula.

7

When the cream is poured slowly into the bowl, it is easy to adjust the amount by checking the scale while removing the air bubbles.

8

Wrap the scaled cream and keep it in the refrigerator.

point This process is called the aging of fresh cream. It is recommended to use the icing cream after it has been aged 30 minutes before use. **31p**

9

Use the remaining cream immediately as an insert cream (for filling) after removing the cream for icing.

point Insert cream is recommended to whip to 90%, but it can be adjusted between 90 to 100% depending on the room temperature and application. In summer, when the temperature and humidity are high, the cream droops easily, so it is stable to whip more than 90% for durability.

10

11

After filling all the insert cream, ice with the icing cream that has been aging in the refrigerator. The optimal density of icing cream is 85%. When the cream is folded with a rubber spatula, lines are formed on both sides of the spatula, and the peak of the cream slightly bends over, and the surface of the folded cream appears smooth- this is the optimal icing density of 85%.

point Working speed is important for icing cream. If icing is done within ten minutes, it is better to whip to 85% density, and if the time goes over 10 minutes, it is recommended to start with the cream whipped thin to 80% density.

To make a cake with a good level of completion, it is better to whip the cream for decorations eparately. Depending on the type of decoration, the amount of whipped cream can be used from 80 g to as much as 150 g. In order to make the decoration cream that is strong, sharp, and does not droop, finish by slowly rotating the mixer at low speed (1st) during the finishing step of evening out the cream.

Reason for aging whipped cream in a refrigerator

The reason for the aging of whipped cream is to maintain the material temperature and workability, and it is good to know the following process.

❶ While working with whipped cream, the temperature must be maintained until the icing is completed in the ice-bearing bowl; this is called 'material temperature,' and the material temperature should be maintained at 3~7°c.

❷ The cream whipped to 70% can maintain the material temperature better than the fresh liquid cream until the icing is completed. Additionally, when the cream is whipped to 70% in advance, the workflow speeds up to the next step (icing), so the aging of the whipped cream is essential.

❸ While the icing cream is aging, fill the sponge sheets with the insert cream.

❹ After the insert cream is filled, whip the aged cream to 80~85% depending on the working speed.

❺ Density of decoration cream should differ according to the nozzles, and even if only a small amount is needed, it is better to make enough cream for easier whipping.

❻ As for the decoration cream, pipe as quickly as possible because the density of the cream in the front of the piping bag and the cream in the back of the bag may be different.

02 Types of cream and appropriate whipping density

Depending on the type of cream, the density and increase rate of volume varies.
The speed of whipping fresh dairy cream is generally 'low speed → medium-high speed → low speed,'
and the whipping speed is lower when the viscosity of the cream is higher. Whip at low speed to
dissolve sugar, and when it reaches the stage where small bubbles appear evenly on the surface, whip
at medium-high speed to increase the volume of the cream lightly, then whip again at low speed
to even out the texture of the cream. The whipping time is very important because the form of the
cream is determined by it. Because the cream will not obtain its form if it's not whipped enough,
and if it's whipped too much, the cream will separate.

Density of cream

Round nozzle < Closed/Open star nozzle < Icing cream < Scooping technique, Ruffle nozzle < Insert cream

← 80% ———————————————— 85% ———————————————— 90~100% →

Thin |||||||||||||||||||||||||||||||||||||||||||||||||||||||||||||||||||||||||||||||||||||||||||||| Thick

Type of cream	Workability			Viscosity		Volume increase rate	
	Easy	Normal	Hard	Low	High	Low	optimal
Mascarpone whipped cream		○			○		○
Compote whipped cream	○				○		○
Fruit puree whipped cream			○	○		○	
Ganache whipped cream			○	○		○	
Cookie base whipped cream		○			○		○
Natural powder whipped cream		○			○	○	

1. 60% density

The cream drops immediately when scooped with a rubber spatula. It is usually used for mousse cakes.

2. 70% density

The cream has fluidity. When scooped and dropped with a rubber spatula, lines form and disappears immediately. This state is the density used to age the icing cream.

3. 80% density

When the cream is scooped up with a spatula, lines are formed on both sides of the spatula, and its peak bends over a lot. It is a suitable density for piping with round nozzles, and I recommend to start icing with a little bit thinner density of 80% for beginners.

4. 85% density

This state is suitable for icing. When the cream is scooped and folded, lines appear on both sides of the blade, and its peak bends over slightly. The surface where the spatula has passed is smooth and glossy.

5. 90% density

This density is suitable for a cream that pipes sharp lines and requires strength, such as a ruffle nozzle. It is also suitable for decoration that uses a spatula for scooping technique. Beyond 90% density, the characteristic refreshing taste of fresh cream turns greasy. Use this as an insert cream to enhance the taste of fresh cream and to maintain the shape of the cake.

6. 100% density

In the cake design class, priority goes to proper hardness, shape maintenance, and retention among the various elements of whipped cream. Usually, 90~100% density cream is used as insert cream in the design class.

Mascarpone whipped cream

In order to give a more rich flavor to fresh dairy cream, 10% mascarpone cheese to the weight of cream can be mixed to make more elastic cream with high viscosity. The cream separates rather quickly, so when making it as in icing cream, it is better to start thinner than the suggested icing density (85%). When making it a decoration cream, be careful not to put too much quantity in a piping bag at once.

Compote whipped cream

A typical example is a cream using blueberry, raspberry, blackberry, and passion-mango compotes. Based on 180 g of icing cream, add 70 g of blueberry compote or blackberry compote, 80 g of raspberry compote or passion-mango compote. For blueberries, smaller fruits have less juice, so it is better to use larger fruits. More ripen fruit has deeper purple color; less ripen fruit has magenta color. Compote whipped cream has more viscosity, so it is relatively easy to do icing due to its elasticity. If the amount of compote added is too little, be careful as the insert cream may bleed on to the surface of the icing cream and may look messy over time.

Fruit puree whipped cream

Raspberry and strawberry puree are favored, and 30% of the weight of the cream is used. Fruit puree whipped cream is most challenging to handle because the cream separates easily due to its high acidity, so it should be used by whipping at low speed.

Cookie-based whipped cream

Generally, Cookie-Crunch (ready-made) can be divided into Oreo, Caramel Cookie, Lotus, Mango Crunch, and Strawberry Crunch. When making cream using commercial Oreo Cookies, a 15 cm cake uses ten pairs of cookies ground in a blender or crushed with a rolling pin (use only the cookies after scraping off the white cream with a spatula). The commercial cookies, such as Oreos, are useful to flavor the cream because of its sweet and salty taste. But if the ground cookies are whipped in the cream from the beginning, the color of the cream will darken and will not look nice. Therefore, it is recommended to use the commercial cookies for the insert cream and use Cookie-Crunch to effectively give black and white contrast for the icing cream on the surface of the cake. It looks best to whip fresh dairy cream first to 75~80%, then mix the finely ground Cookie-Crunch bits with a rubber spatula. If the density is thin, use a whisk or hand mixer at low speed to adjust.

Ganache whipped cream

Ganache whipped cream is one of the tricky creams to deal with. Make the ganache by mixing chocolate and fresh cream to a 1:1 ratio. Add this ganache to fresh cream whipped to 60% and mix evenly with a rubber spatula or a whisk. Ganache whipped cream separates easily, and when used in an ice bath, it starts to harden from the bottom. Therefore, it should be removed from the ice bath as soon as the cream is made and use immediately.

Natural powder whipped cream

Natural powder is relatively easy to handle depending on the type, but some are difficult to handle, such as strawberry powder and raspberry powder, because they lump easily. Sweet pumpkin, mugwort, and sweet purple potato powder are relatively easy to use when whipping. For the natural powder, it is recommended to use 10 g for 180 g fresh cream for icing. This is because if more powder is used, the cream becomes heavy with high viscosity, and the volume increase rate is low, making it difficult to ice with. Just as the powder does not usually dissolve well in cold water, it does not dissolve well when the temperature of the cream is too low. Therefore, take the bowl out of the ice bath and sieve the powder into the fresh cream and whip it at low speed, cleaning the sides alternately with a rubber spatula. When all the powder is dissolved, place it back on the ice bath and whip it at medium speed to make an appropriate density. Natural powder whipped cream is prone to become dry if it's whipped too much, so be careful not to increase the whipping above medium speed.

Cutting Sponge Cakes and Filling with Whipped Cream

This is
the preparation process
before going into actual icing.
Depending on the cake, the method of
cutting the sponge and filling the whipped
cream differs by whether fruits are included
between the sheets or not. This chapter
explains, based on the meticulously
organized Congmom's know-how, from
the class experiences so far that
even beginners can easily
understand.

01

Before cutting and filling

Before cutting the sponge by the the shape of the cake and filling the whipped cream,
this is how to sort the whipped cream according to its purpose and how to fill the cream.

: Classification of cream according to its use

Icing cream

The cream that is used to cover the cake is called 'icing cream.' It is used to ice the cake, and it can be expressed in various colors and textures by mixing compote, sprinkles, etc., to create the overall atmosphere of the cake.

Decoration cream

The cream that decorates the iced cream using various nozzles is called 'decoration cream.' The cake can be completed in various designs depending on the nozzles and the techniques used.

Insert cream

The cream used in between the cake sheets is called 'insert cream.' The amount of cream used, and the thickness of the sponge sheets vary depending on whether fruits are added or only the cream is filled.

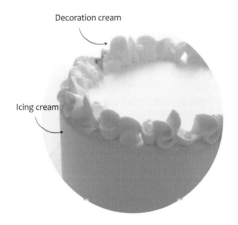

Decoration cream

Icing cream

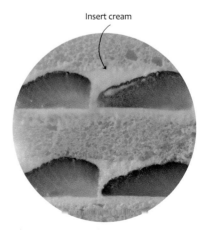

Insert cream

: Cutting sponge cake and filling cream depending on the use of fruits

How the sponge is cut and filled will vary depending on whether the insert cream contains fruits.

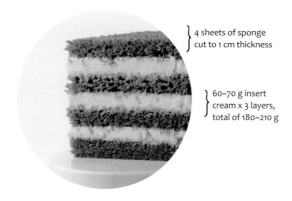

4 sheets of sponge
cut to 1 cm thickness

60~70 g insert
cream x 3 layers,
total of 180~210 g

When the cake is filled only with cream, the sponge cake is usually cut into four sheets of 1 cm thickness. But for a special custom cake, five sheets of 1 cm thickness are used for extra height.

For a 15 cm cake, 60~70 g insert cream is used per layer, which totals to 180~210 g, 160~170 g as icing cream, and 80~150 g for decoration cream.

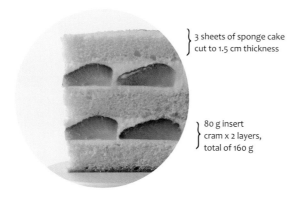

3 sheets of sponge cake
cut to 1.5 cm thickness

80 g insert
cram x 2 layers,
total of 160 g

When filling with fruits, it's common to use three sheets of 1.5 cm thickness sponge sheets. When the middle sheet is cut to 1 cm thickness, the texture in the mouth changes, and depending on the preference, only the middle sheet can be cut thin.

80 g of insert cream is used per layer, which totals to 160 g, 180~190 g as icing cream, and 80~150 g for decoration cream.

The amount of icing cream is larger than that of a cake filled only with cream is because when fruits are added, it increases the height. The amount of insert cream and icing cream can be adjusted within 10 g depending on the height of the fruits being added.

02

Cutting and filling round cake 1

It is a method of cutting a 15 cm sponge cake into four sheets of 1 cm thickness and filling only with the cream. It is important to fill each layer with an equal amount of insert cream. Here is how to fill effectively.

1

Using 1 cm thick confectionery bars and a bread knife, cut and prepare four sheets of 1 cm thick sponge sheets.

2

Whip 160 g of insert cream. Set the hand mixer to 2nd speed and increase the volume of the cream.

3

The blades of the hand mixer may not reach the cream on the edge of the bowl, and density may be thinner than the cream in the middle of the bowl. From time to time, while whipping, use a rubber spatula to scrape the cream from the edge of the bowl to even out the overall density.

4

When the cream is scooped with a rubber spatula, check if the peak of the cream does not fold back and retain its shape.

5

Check the density of the cream, and reduce the speed of the hand mixer to 1st speed to even out the cream texture, and adjust the density of the insert cream to 90% or above.

6

Organize the texture of the cream once again with a rubber spatula and level out the cream.

7

Hold the edge of the rubber spatula straight and divide the cream into three equal parts. This is a preparation process so that the same amount of cream can be filled per layer.

8

Place the bottom sheet of the cake in the center of the turntable, and scoop up all the cream from one of the three equal portions and place it on the sheet.

9

Spread out the cream with a spatula while turning the turntable.

10

After spreading, drag the cream inward along the edge of the sheet.

11

Spread the top with the spatula while taking care not to let the insert cream go out of the sheet.

12

Hold the spatula vertically and rotate the turntable to organize the side. At this time, don't push the spatula hard to prevent from making too much crumb.

13

Neatly trim the taller cream on the side.

14

The taller cream must be trimmed neatly so that the edges can be aligned well when placing the second sheet.

15

Look down from the top and make sure to align the edges of the first sheet and the second sheet.

16

After placing the second sheet, shift the sheets up and down using both thumbs and align the edges of the two sheets.

17

Push the protruding part in with your thumb, pull out the edge that is too far in, and accurately align the edge.

18

Repeat to fill the same way until the fourth sheet is placed aligning the egdes. Organize the sides using a spatula while turning the turntable at the same time to finish neatly.

03

Cutting and filling round cake 2

This is a method of cutting a 15 cm cake into three sheets of 1.5 cm thickness, then placing fruits on every layer with cream. As with the cake filled only with cream, it is important to fill with the same amount of cream on each layer. As a filling method used most often, here is the know-how that Congmom has organized after various attempts.

1

Using 1.5 cm thick confectionery bars and a bread knife, cut and prepare three sheets of 1.5 cm thick sponge sheets.

2

Whip 160 g of insert cream to 90% or more and level out the cream. Divide the cream into four equal parts with the blade of the rubber spatula.

3

Place the bottom sheet of the cake in the center of the turntable.

4

Scoop out one section of the divided cream. Leave a little bit of cream from the section, and place the rest on the sheet. Remember, when fruits are filled, that the cream spread under the cream is less than the cream applied over the fruit.

5

Spread out with a spatula while turning the turntable.

6

After spreading, bring in the cream from the edge to the center, filling the cream within the edge.

7

Spread the top evenly while watching the cream not go over the edge too much.

8

Place the fruits about 0.5 cm in from the edge of the sheet. If the fruits are placed too far from the edge of the sheet, the edge can gradually sag and form a dome.

9

Push the fruits lightly with fingertips to prevent the fruits from sliding.

10

First, drop the cream that's left on the spatula on the fruits.

11

Scoop the remaining cream from procedure 4 and another whole section of the cream on the fruits.

12

In order to cover the fruits neatly at once, the fruits should be covered with a larger amount than the cream under the fruit.

13

Now, only half of the cream is left in the bowl.

14

Spread the cream evenly to cover the fruits.

15

Hold the spatula vertically and rotate the turntable to organize the side neatly.

16

Flatten the top to easily align the sides of the second sheet when placed on top.

17

Look down from the top and make sure to align the edges of the first sheet and the second sheet.

18

After placing the second sheet, shift the sheets up and down using both thumbs and align the edges of the two sheets.

19

Push the protruding part in with your thumb, pull out the edge that is too far in, and accurately align the edge.

20

Fill the cream and fruits the same way and place the third sheet, aligning the sides precisely.

21

Neatly organize the sides once again with a spatula while turning the turntable.

04

Cutting and filling dome cake

Sponge sheets used for dome cake are cut to 3 sheets of 1.5 cm thickness. One of the three sheets, which will be the top of the cake, must be cut smaller than the remaining two to make the dome-shaped cake. There are two ways to cut the top sheet; using kitchen scissors and cutting with a mousse ring. When cutting with a mousse ring, the sharp corner can be seen after icing, so it is recommended to trim the corners with kitchen scissors.

1

Using 1.5 cm thick confectionery bars and a bread knife, cut and prepare three 1.5 cm thick sponge sheets.

2

Stack all three sheets in the center of the turntable.

3

Place left hand lightly on top of the sheets and trim the corner of the very top sheet with kitchen scissors while turning the turntable.

4

Trim it to form a rounded dome.

5

Fill the insert cream the same way as in '03. Cutting and filling round cake 2', and place the top sheet trimmed round. `44p`

6

After placing the second sheet, shift the sheets up and down using both thumbs and align the edges of the two sheets.

7

At this time, push the protruding sheet inside using the knuckle of the thumb. That way, the top sheet won't stick out.

8

Form the dome shape by pressing with your thumb.

9

Hold the spatula lightly and start organizing the sides of the top by keeping the spatula aligned diagonally with the dome shape.

10

Now, hold the spatula vertically and organize the sides of the bottom of the cake.

11

Repeat procedures 9 and 10 to finish filling the dome cake.

Cutting the sponge cake and filling it with whipped cream is very important, depending on the presence of fruit to fill. This was covered in '01. Before cutting and filling', but let's review it once more with a table.

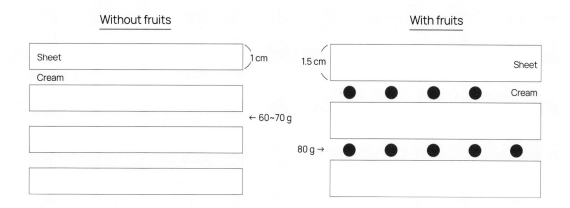

Types of cream	Without fruits	With fruits
Insert cream	(60~70 g) × 3 layers = 180~210 g	80 g × 2 layers = 160 g
Icing cream	160~170 g	180~190 g
Decoration cream	80~150 g	80~150 g

LESSON 04

Icing

This chapter will explain icing techniques according to the various shapes of cakes, from the most basic round cake to dome, chiffon, square, and heart. It also includes how to decorate the sides of a dome cake in different ways using dome scrapers.

01

Traditional technique of icing round cake

8-inch and 9-inch spatulas are used for icing the basic round cake by the traditional technique. 9-inch is used to organize the bottom and lift and transfer the cake, and an 8-inch spatula is used for all the other procedures.

{ Density **85%** | Skill ●●●●○ }

1

Evenly organize the texture of the cream with a rubber spatula.

point The reason for arranging the texture of the cream is to even out the air bubbles created by a hand mixer or a whisk and to set the appropriate density for icing.

2

Top with enough cream, so it flows down about 1/3 the height of the side on the cake.

3

Move the rubber spatula to the back of the bowl. That way, it's comfortable to use the front of the bowl.

4

Divide the top of the cake into four, number the sections, and find the center.

Outer end of the blade

Inner end of the blade

5

The end of the spatula is not the highest end, but both ends of the blade.

6

When pushing out the cream, place the inner end of the blade so that it is on the center of the cake, then push the cream out towards the edge.

7

When pulling in the cream, drag from the outside of the cake towards the center so that the outer end of the blade is at the center.

8

When you repeatedly push and pull the cream while rotating the turntable, the shape of petals will appear on the top of the cake.

9

The more you spread the cream in and out repeatedly, the spacing of the petals will become tight. If so, the texture of the cream may turn rough, so it is best to have five petals appear on a 15 cm cake.

point At this stage, it is important to let the cream flow down 1/3 the height of the side of the cake. This is because that cream plays a role in building a crown.

10

Place the outer end of the blade on the center of the cake, the rear blade touching the top of the cake, and slightly open the angle to level to the top while turning the turntable.

point Make sure the right-hand does not move at this time. To make sure the cream is evenly leveled, your eyes should be watching where the spatula has passed. It is easier to work by looking at the center point of the cake because lines can appear on top if the eyes are watching elsewhere, even for a second.

11

Place just the top part of the blade on the left side of the cake (between 1 and 3), and lightly spread the cream, building a crown around the top.

12

Use the remaining cream in the bowl to ice the side of the cake.

13

Spread the cream evenly on the side of the cake while moving the spatula back and forth in big movement and adjusting the thickness.

14

Clean the messy cream on the bottom by lightly sweeping with the spatula.

15

Hold the spatula vertically on the left side of the cake (between 1 and 3).

16

Make the rear blade of the spatula adhere to the side of the cake, open the front blade slightly, and ice the side neatly by turning only the turntable without moving the right hand.

point At this time, the moment your eyes are turned behind the spatula to check where it has iced, the posture can shift, and the shape of the cake may change. Ice the cake, with your eyes watching the front of the spatula, checking that the cream is being filled, and then make sure that the cake is perpendicular overall.

17

Adhere a 9-inch spatula on the turntable.

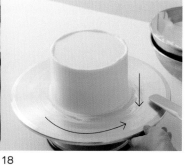

18

While rotating the turntable, pull the spatula inward to clean the surface.

19

From section 2, trim the crown to level the top.

<u>point</u> ❶ At this time, a sharp edge can be made by opening the angle of the spatula. Open the angle when the crown is high, and close when it's low to level the cake.

❷ Be careful not to let the front blade go out of the section to be leveled.

20

When the top is leveled, hold the spatula long and horizontally, and finish up by lightly passing across the top from edge to center.

<u>point</u> To let the hand relax, hold the handle of the spatula, not the neck.

Tip

One of the critical points when icing is that the cream should not come up on the blade. The reason is that the cream that got on the blade may drop elsewhere and become dirty. It is important to scoop while controlling the right amount of cream.

02

Simple technique of icing round cake

This is a technique even beginners who are not familiar with icing can apply easily and quickly.
Use a spatula the same way as the traditional icing technique from procedures 1~14.
After that, the key is to use an angled scraper instead of a spatula.

{ Density 85% | Skill ●●●○○ }

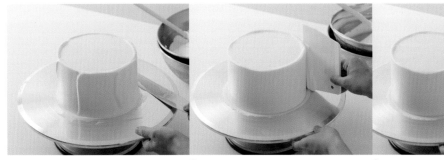

1

Apply the same method as procedures 1~14 from '01. The traditional technique of icing round cake.' `54p`

2

Use a scraper instead of a spatula to finish the side.

point The most significant advantage of icing using a scraper is that the bottom can be finished neatly and that the crown can be built while icing the side at the same time. Taking advantage of being able to organize the bottom neatly, icing can be done on the cake board for more complete icing.

3

When finishing with a scraper, it is most important to keep it vertical.

point ❶ It is convenient to use a scraper with a right angle, which is easy to align vertically.

❷ Using a scraper reduces icing time than the spatula and allows a high degree of completion and clean finish.

4

Use a thin and sharp scraper and run it along the surface to clean any cream left on the turntable.

5

Cream on the bottom is neatly arranged.

6

Wrap the blade of a spatula with a wet tissue, and wipe the bottom clean once again.

7

Starting from section 2, pull the cream of the crown inward to finish icing.

8

When the top is leveled, hold the spatula long and horizontally, and sweep to finalize.

03

Icing dome-shaped cake

03-1 Icing dome cake- Basic

Dome scraper is used to ice dome-shaped cakes. If the dome scraper is not available, cut a thick plastic sheet and bend it to use. Commercially available dome scrapers are mainly divided into three shapes. Scraper with no design is called straight edge or plain edge dome scraper, the one that has a pattern of narrowest intervals is called hair-line or thin wave-patterned dome scraper, and the one that has thick patterns are called thick-line or thick wave-patterned dome scraper. There is a trend in dome scraper as well. These days, a simple plain edged dome scraper is most loved.

{ Density 85% | Skill ●●●○○ }

[Preparation]

① Commercially sold dome scrapers have a narrow and bent angle. If it's used as is, the cake may turn out to be a flat dome with its height too low, or the scraper may excessively dig the bottom of the cake, resulting in an unstable reversed dome shape.

② Therefore, when using the dome scraper, it is best to open the angle according to the height and the form of the cake before using it.

③ This is how it looks when the angle has been opened to the height of the cake.

1

Evenly organize the texture of the cream in the bowl with a rubber spatula, and top with 2/3 of the cream on the cake.

2

Spread the cream evenly with a spatula while turning the turntable.

3

Make the spatula perpendicular to the turntable, the tip touching the plate, spread the cream evenly while turning the turntable.

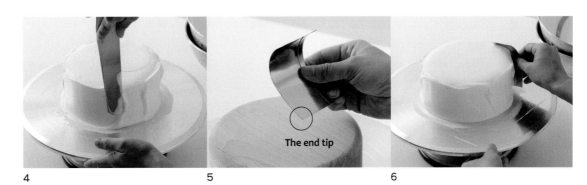

The end tip

4

Fill with the remaining cream as needed.

5

When using the dome scraper, relax the hand and hold the edge lightly, open the angle from the surface to about 15~20 degrees, and apply force to the end tip.

point While working with the dome scraper, it is important to ice without opening the angle of the scraper too much. This is because the cream scrapes off as the angle of the scraper opens. Ice the cake with the angle of the scraper opened just enough to fit a thumb so that the cream is applied as the scraper passes.

6

Start icing the side using the dome scraper.

point The key point of icing with a dome scraper is 'moving the whipped cream.' When the scraper touches the surface of the cake, the cream on top moves down, and the cream on the bottom rides up, filling the center space convexly and completing the shape of the dome.

7

Ice until the top diameter narrows, and the crown rises evenly.

8

When you completed icing the side, pull the scraper forward in short movement.

9

Clean the bottom of the dome with a spatula.

10

The dome has a smaller diameter on top than the basic round cake, so be careful not to use the blade too long lengthwise when trimming the crown.

11

Use only the front part of the blade, and work a little bit at a time for a well-rounded finish.

point A dome-shaped cake can get lower in height if the crown is trimmed, so it is also a good idea to decorate with the crown left as is. It can be made into a cute dome-shaped cake with the crown in place by filling it with ganache. On the other hand, if decorating with nozzles, it's better to trim the crown neatly.

When icing dome cakes, be aware of 'reversed dome' and 'flat dome.' To not make an unstable reversed dome cake that the top is larger than the bottom, make sure not to apply excessive force on the end tip of the card

To avoid making a flat dome, make sure to spread open the dome scraper when icing. If the iced top gets too tall, the crown can cave over and curl backward. In this case, spread the top once more to lower the height and finish with the dome scraper.

03

03-2 Icing dome cake- Wave pattern

Various designs can be made with a plain edged dome scraper. With the strength of the hand applied only on the edge of the scraper while the back blade of the scraper adhered to the side of the cake, you can make a wave pattern can be made by rotating the turntable while lightly moving the scraper in and out. The more exaggerated shape can be made by icing with force applied more on the top and bottom.

{ **Density** 85% | **Skill** ●●●●○ }

1
Ice the side as the same method as in '03-1. Icing dome cake- basic.' 62p

2
Prepare to ice by holding the edge of the scraper lightly and placing the card vertically at 3 o'clock on the cake.

3
Relax the hand, adhere the edge of the blade on the cream, and rotate the turntable while waving the scraper.

point By adjusting the speed of hand movement, the width of the wave can get wider or narrower. Move the scraper left and right, not up and down, by maintaining the speed of movement and making a wave pattern.

4

When the wave pattern is complete, pull the scraper out in short movement.

5

When finished, clean the bottom with a stiff scraper instead of a spatula to prevent from ruining the waves.

<u>point</u> If you use a spatula to clean, the wave pattern on the bottom disappears. Make sure to use a scraper to maintain the pattern. To prevent the scraper from touching the waved cream, place it at about 45 degrees and clean each curve of the wave.

6

Repeatedly wipe off excess cream on the back blade of the scraper while cleaning up.

7

The cake is neatly cleaned with the wave pattern on the bottom visible.

03

03-3 Icing dome cake- Modified wave pattern

This design is used when a simpler design is wanted than the wavier pattern.
The method is to stamp one by one, with the hand relaxed. I recommend this technique for
those who have difficulty icing the wave pattern, which requires continuous motion.

{ **Density** 85% | **Skill** ●●●○○ }

1

Ice the side as the same method as in '03-1. Icing dome cake- basic.' **62p**

2

The pattern will be made by repeating the process of dipping and pulling out from the cream using the plain edged dome scraper. Hold the dome scraper stably with all five fingers to evenly distribute strength, and start icing while holding the scraper vertically.

3

While rotating the turntable, take care not to put excessive strength to the hand holding the scraper, and repeat dipping and pulling to make patterns. Stamping in a regular pattern finishes beautifully as well.

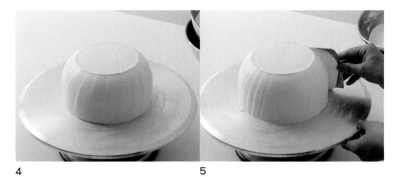

4

The bottom is neatly cleaned.

5

By adjusting the spacing, it can be improvised into the desired design.

03

03-4 Icing dome cake- Narrow lines

Using a dome scraper with a ripple design creates narrow lines.
Narrow line design can be used to make a tambourine cake.

{ **Density** 85% │ **Skill** ●●●○○ }

1

Ice the side by applying the same method as in '03-1. Icing dome cake- basic' with a ripple edged dome scraper. `62p`

2

When the narrow lines are completed, clean the bottom with a spatula.

3

The cake is finished neatly.

03

03-5 Icing dome cake- Wide lines

By using a thick wave edged dome scraper, you can make an elegantly designed cake.
It can be made into Igloo cake during the winter season.

Density 85% | Skill ●●●○○

1

Ice the side as the same method as in '03-1. Icing dome cake- basic' with a thick wave edged dome scraper. 62p

2

When the wide lines are completed, clean the bottom with a spatula.

3

The cake is finished neatly.

04

Traditional technique of icing chiffon cake

Chiffon cake icing has many of the same processes as icing round cake, but the difference
is that the hole in the center gets iced and trimming the crown of the hole.
Be careful that if a large amount of cream is iced on top,
the whole cake may droop down and turn into a dome shape.

{ Density **85%** | Skill ●●●●● }

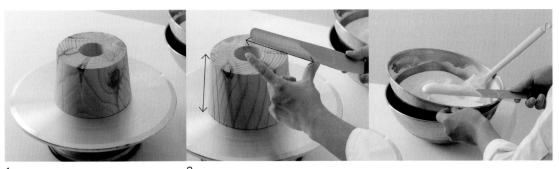

1

2

The wooden chiffon model used
here is custom made into the size
of a Japanese chiffon cake pan, in
the size of 15 cm diameter at the
bottom, 13 cm diameter on the
top, 10 cm in height, and 4.5 cm
diameter hole in the center.

Scoop the cream as long as the height of the cake with a spatula.

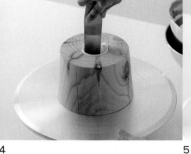

3

Start pre-icing thinly from the right inside the hole, with a spatula.

4

Spread the cream lightly once without going back and forth several times to prevent making too much crumbs from the cake.

point Do not reuse the cream on the spatula after icing. It's because for chiffons, as the spatula contacts the cake, it continues to make crumbs. When icing the cakes, always put a small bowl aside and scrape off the cream from the spatula.

5

When the pre-icing is finished, gently pull the spatula out of the hole with the turntable spinning while supporting the blade with the left index finger so that the spatula does not touch other parts.

point If the spatula is removed while the turntable is stopped, a line may remain inside the hole. Also, if the spatula is pulled out with only one hand, it can touch the hole; therefore, remove the spatula with both hands while the turntable is spinning.

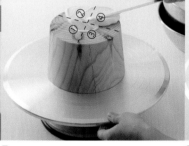

6

Scoop the cream with the spatula, wider than the radius of the top.

7

Spread the cream by pulling the cream twice in a clockwise direction from section 4. The more the cream is spread, the more crumbs appear, so pull in only two times.

point At this time, adjust the length of the blade to spread the cream so that the cream drops both inside the hole and outside the cake. The dropped cream is used to build a crown in the hole and on the outer edge of the cake.

8

Level the cream flat by rotating the turntable while keeping the spatula at section 4.

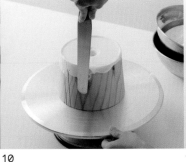

9

Starting from the side of the cake (center of sections 1 and 3), place the spatula so that only the top part touches the cream and build the crown.

10

Same as before, build the crown in one movement without moving the spatula back and forth to prevent crumbs from forming.

11

Scoop the cream with the spatula and start spreading from the bottom. Likewise, to prevent crumbs from forming spread the cream back and forth only twice.

12

Because chiffon cake is high, there may be areas on the side that isn't iced.

13

Use the spatula to pull the cream up from the bottom to top, to cover all sides.

14

Place the spatula diagonally, adjust the cream on the side in even thickness, and check to make sure the crown is built evenly.

15

Neatly clean the surface of the turntable with a spatula.

16

Start from the center of the left side of the cake, finish icing with the rear blade touching the side, and the front blade slightly open.

17

Clean the bottom without putting the spatula deep underneath the cake.

point Be careful not to put the spatula too deep into the bottom of the cake to avoid creating a gap.

18

Place the spatula on the wall of the hole, in the direction of section 2, and rotate the turntable to build the inner crown.

point Depending on the type of the chiffon cake pan, the shape of the hole may differ. If the diameter of the hole becomes wider from bottom to top, the spatula has to be tilted as well towards the outside rather than vertical. Adjust the angle to fit the pan so that the cake in the hole does not show through the cream.

19

When the crown is up and the icing is done inside the hole, with the turntable spinning, support the spatula with the left index finger, and pull it out from the hole.

point If the spatula is removed while the turntable is stopped, a line may remain inside the hole. Also, if the spatula is pulled out with only one hand, it can touch the hole; therefore, remove the spatula with both hands while the turntable is spinning.

20

Hold the spatula on the top of the cake from the direction of section 2, and trim the crown.

point Be careful not to touch the inner crown.

21

Trim the inner crown from the direction of section 3.

22

Hold the spatula short and level the top.

23

The cake is finished neatly.

Tip

It is important to control the amount of cream when icing chiffon cake. Because the cake itself is light, if too much cream is iced on top, the cake may sag towards the center hole of the cake. Also, due to the weight of the cream, the top part of the cake and the cream may sink down, and the whole cake can collapse into a dome shape. Therefore, it is important not to give excessive force on the hand, and not to use too much cream at once.

It is also important to trim the crown while keeping the hole in the center of the chiffon in a circular shape. Trim with a spatula by touching a little bit at a time.

05

Icing square cake

When icing a square cake that has angles, using a scraper than a spatula makes the cake finished cleaner and can shorten the working time.

Density 85~90% | Skill ●●●●○

1

Divide the top of the cake into four sections, and number them.

2

Top with cream enough to flow down till 1/3 of the side.

point If too much cream is topped, the cream keeps flowing down, making it difficult to build a crown. The same applies when icing shaped cakes.

3

Start at the center of sections 3 and 4 and spread out the cream by pushing outward twice.

point At this time, let the cream flow down the front and side of the cake.

4

Push and spread the cream into four equal parts while rotating the turntable.

5

The cream is spread out into four.

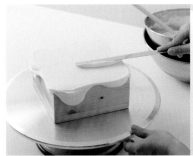

6

Starting from section 4, spread the cream to flatten the top while rotating the turntable.

7

Starting from the side of the cake, place the spatula so that only the top part touches the cream, and build the crown.

8

Build the crown evenly on all sides.

point To make the corners stand out, cut off icing at each corner instead of moving the spatula in connected motion.

9

Use the remaining cream in the bowl to finish icing the sides.

point The reason why icing square cake is difficult is that all four corners should stand out nicely. If there is a part where the crown is not clearly raised, use the restoration tip to build the crown quickly. 94p

10

Place the scraper on the right side and ice all four sides evenly. Be careful not to open the angle of the scraper too much.

11

Clean the bottom with the spatula.

12

Hold the spatula short, and place the spatula on the corner of the cake.

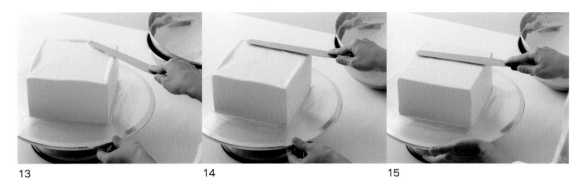

13

Start trimming the crown from the corner.

14

Trim the side as well.

15

Repeat trimming the corner and side in order, and level the top to finish.

06

Icing heart-shaped cake

Just like the square cake, a heart-shaped cake can also be iced using a scraper, making the cake finish cleaner and shortening the working time.

Density 85~90% | Skill ●●●●○

1

Divide the cake into the left and right sections.

2

Top with cream enough to flow down till 1/3 of the side.

3

Start icing from the right side of the cake. Push the cream out from the center towards the edge.

4

Spread the cream side to side from the top towards the bottom, and fill the right side of the cake.

point The cream should reach the edge of the heart to be able to build a crown.

5

Spin the turntable and ice the left side using the same method.

6

Level the top by keeping the angle of the spatula open while rotating the turntable.

7

Build the crown starting from the corner.

8

Keep building the crown while rotating the turntable.

9

Continue icing the sides, filling cream as needed.

10

Clean the surface of the turntable.

11

Set the scraper on the curved side of the heart as a starting point.

12

Rotate the turntable wide with the left hand. Hold the scraper vertically with the right hand and turn it to the bottom corner of the heart in one sweep.

point Unlike the previous icing method using a scraper shown earlier, be careful not to close the angle of the scraper too much. It should feel like sweeping lightly with the edge of the scraper in order to make the shape.

13

Do not continue icing, but pull out the scraper straight from the corner below the heart.

14

Rotate the turntable and start icing again from the corner of the heart.

15

In the same way, icing will be done in one movement towards the center of the curved side.

16

While opening the angle of the scraper, rotate smoothly and stop at the center of the curved surface.

17

Face the curved side of the heart and ice it towards the right.

18

Clean the bottom with a scraper.

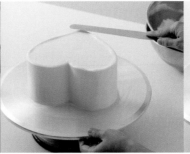

19

Wipe the surface with a spatula wrapped with a wet tissue.

20

Trim the crown a little bit at a time, starting from the corner of the cake, holding the spatula short.

21

For the shaped cakes, it is better to trim the top in several little strokes. That way, it can maintain its shape and form better.

22

Once the top of the cake is trimmed, hold the spatula long, and finish by sweeping the top in one stroke.

Comparing Korean,
Taiwanese and
French Scrapers

Korean   Taiwanese   French

- Because the Korean scraper has a diagonal shape, it is difficult to ice vertically, and the side that touches the cake is narrow, so there is a disadvantage of having to turn the cake a lot to ice evenly. On the other hand, the bottom side of the Taiwanese card is horizontal, so it is easy to set the scraper vertically. Additionally, the side that touches the cake is gradual and has a wide width, which minimizes the number of rotations and keeps the cream in good condition. French scraper is stiff and has sharp edges, which is suitable for cleaning the surface of the turntable.

- When using a Taiwanese scraper, place it on the right side of the cake at a 3 o'clock location, give force on the lower point of the scraper, and ice. This is to neatly finish the bottom and enhance the completeness of the icing. (If it's difficult to align the side vertically, sit in a chair and work at eye level.) Also, make sure to hold the bottom of the scraper so that the side can be iced vertically, not diagonally.

How to Restore
and Finish Icing

When icing, the shape of
the cake or the consistency of the cream
can change and may need to restore it. This
chapter will cover how to correct the shape
of the icing and restore separated cream,
how to transfer the cake to the board
safely, and how to finish neatly.

How to restore and finish icing

Whipped fresh cream is sensitive to temperature and density, which there may be cases that may need to be restored after icing. The cases in which it will be most likely to fail are when the shape of the side of the cake is not straight, the cream is separated, the crown is not raised correctly, or the bottom is uneven. But don't worry. Here is Congmom's know-how of each recovery tip in detail.

(01) When the side is pyramid shaped

Often during the class, some habitually finish the icing in a pyramid shape. The reason why it is completed in a pyramid shape is that the spatula is not placed on the center of the side but in the back. When the spatula touches the back, the force is applied on the handle, and it is inevitable to ice diagonally. To restore from the shape of the pyramid, the large amount of cream at the bottom should be pulled up in circular motion and align the side vertically. This method is also used to ice taller cakes.

[Incorrect position]   [Correct position]

1

The cake is incorrectly completed, which the sides are iced diagonally.

2

The correct position to start icing is the center, not the back.

3

Use a spatula to pull up the cream that has collected a lot on the bottom.

4

At this time, the key is to move the whipped cream upward by pulling up from the bottom in a circular motion.

(02) When the side is inverted-pyramid shaped

This is the opposite case of completing in a pyramid shape. Mainly, the spatula is placed in the front rather than the center of the side so that the angle gradually opens, and then it becomes an inverted-pyramid shape. Make sure to ice the side at the center of the left side of the cake.

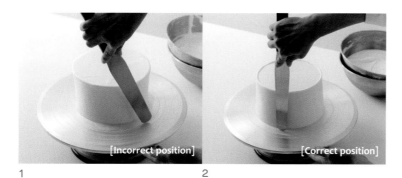

1

When icing the side, there are times where the upper angle of the spatula opens as the spatula gradually moves forward.

2

The position of icing the side is the left side of the cake from the center. Always remember this position while icing.

(03) When the cream on the side is separated

If the cream on the side is separated after icing is completed, heat the spatula on a blow torch, cool it for a while, then use the residual heat to re-ice the side of the cake to restore. This method is recommended for experienced people because the spatula cools down before half a turn of the cake is done, so it must be reheated again with the blow torch continuously. Here are two other ways that even beginners can easily recover at once.

[By using hot water]

1

Dip the spatula blade in hot water for a while.

2

Dry the blade thoroughly with a dry cloth.

3

Use the residual heat of the spatula to restore the cream by icing the side of the cake again.

point Because of its heat persistence, it can make one full turn.

[By using liquid fresh cream]

1

Drop 3~4 drops of fresh liquid cream on a spatula.

2

Spread the cream evenly on the spatula with gloves on.

3

Ice the side of the cake again with the spatula with the liquid cream.

point This is the most convenient way, to rotate the cake two turns with a spatula moistened with liquid cream. If too much liquid cream is used, it can make the bottom of the cake watery, so adjust the amount carefully.

(04) When the entire cream on the cake is separated

If the entire cream is separated after the icing is finished, mix just a small amount of the fresh liquid cream with the remaining whipped cream (based on 100 g whipped cream, add 10 g liquid cream), apply a thin layer on the entire cake, then ice again. It is better to use a rubber spatula when mixing. It's because if a hand mixer is used, the density of the cream may suddenly increase. If too much liquid cream is added, the bottom of the cake may become watery, so the quantity should be adjusted to recover. The density of the cream to be used for restoration should be thinner than the cream that's been iced with. When the density is thinner, it will become an appropriate density of 85% when it contacts with the separated cream.

Overall, if the cream seems to be frequently separating, check if the time it took to ice wasn't too long. Even if the cream is at an appropriate density to start with, the cream will separate after 10 minutes, so it's good to practice to finish icing within 10 minutes after setting a timer.

(05) Simple method to build the crown on the cake

Sometimes when icing, the crown does not seem to rise evenly. The main reason the crown does not form is that the spatula did not adhere to the side of the cake. If the crown does not rise clearly, here is a simple way to do it.

1

Place the spatula diagonally at the back of the side of the cake (11 o'clock location), and let only the top part touch the cake.

2

With the spatula angle open, rotate the turntable to build the crown.

<u>point</u> At this time, make sure not to move the spatula back and forth to ice. While rotating the turntable, build the crown in one try without moving the spatula.

06 When the crown on the cake is partially not built

When icing, there are times where the crown may be missing one or two spots. In such cases, it is more convenient to restore the part that is missing rather than re-icing it from the beginning. Here is how to do it.

1

This is when a part of the crown is missing.

2

Hold the spatula vertically and scoop just a little bit of cream with the tip.

3

Hold the spatula vertically where it needs to be patched.

4

Dip the spatula vertically and top the small amount of cream.

5

When the cream is in place, pull the spatula towards the back to finish restoring.

(07) When the bottom of the cake is not clean

Even if the bottom is being cleaned by attaching the spatula to the surface of the turntable, the cream may still build up and become dirty as the icing time increases. If so, it can be cleaned by opening the angle of the spatula and clean slowly. Note that if the angle is opened too much, the cake sponge may become visible on the bottom.

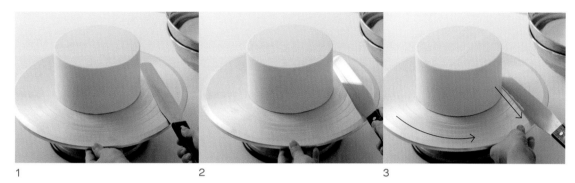

1

The correct posture for cleaning the bottom plate is attaching the spatula to the turntable. Now rotate the turntable to do the first clean up.

2

After the first clean up, if the foot of the cake is uneven or some spots of cream are left behind slightly open the angle of the spatula.

3

In this state, slowly rotate the turntable and, at the same time, pull out the spatula to finish.

4

The bottom is cleaned up neatly.

(08) Transferring and finishing the cake

When lifting the cake with a spatula, two 9-inch spatulas are needed that can go deep and support the weight. However, it may be challenging to lift chiffon cake in which the sponge is weak or tall cake heavy with sponges, so these are better to ice directly on the cake board. Likewise, the cakes that are iced with scrapers are recommended to ice directly on the cake board to take advantage of being able to work the bottom neatly.

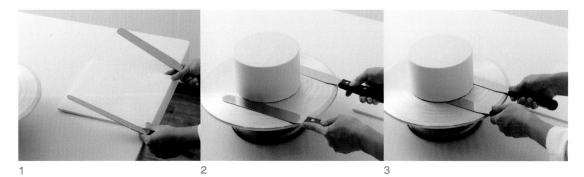

1

When moving the cake, place the cake board at the end of the worktable so that the edge aligns.

2

Find the front of the cake, align it to face the front, and lift the cake from outside using two 9-inch spatulas.

3

If the blades are inserted too deep, the handle of the spatula may go above the turntable, so insert with caution.

4

Lift the cake slowly.

5

Transfer the cake carefully and place it on the cake board.

6

Move the board to the front of the worktable to remove the spatulas.

7

Hold the board and the spatula simultaneously with the left hand, hold only the board with the right hand, and then lift it to a height where it is easy to remove the spatula.

8

If both spatulas are removed simultaneously by spreading them out, the foot of the cake may widen and become a pyramid shape, so turn towards the back one spatula at a time to remove.

<u>point</u> Removing the spatula by turning it around to the back cleans the bottom once more, and the front of the cake can be preserved.

9

The other spatula will be removed the same way. Hold the cake board with the right hand.

10

With the left hand, turn the spatula backward and remove it.

11

After transferring the cake, clean the turntable with a wet tissue.

12

Place the cake board on the turntable, and clean the board with a spatula wrapped with a wet tissue.

The reason why the crown is not easy to build

There are many reasons why the crown does not rice clearly.

First is the issue of the density of the whipped cream. If the density is too thin, it will not have enough strength to hold itself. Also, if the whipped cream is too thin, there can be cases where the sponge may become visible all around after icing. Second, as previously mentioned, the spatula was not in close contact with the surface of the cake. The third is when a large amount of whipped cream was topped on the cake. If the cream is too heavy that it sags down gradually, it is difficult to build a crown. Therefore, it is necessary to practice making the right density and using the right amount.

LESSON 06

Decorating with
Piping Nozzles

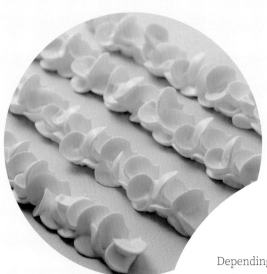

Depending on the type of nozzles and the techniques used, you can complete the design diversely. The most widely used closed star nozzles, round nozzles, and French star nozzles to leaf nozzles, petal nozzles, and drop flower nozzles can create more splendid and unique designs; this chapter will cover various techniques using piping nozzles.

01

Closed star nozzle

When it is the first time piping whipped cream, it's good to start with a closed star nozzle.
It is because a closed star nozzle is the fastest tool to improve piping skills
in a short period of time. It's good to practice with a small nozzle that pipes more precisely
while checking the shape than a larger nozzle. When using a closed star nozzle
for cake decoration, a nozzle with six point or more is recommended.
The more points it has, there are more areas that create shades, which make it look fancier.
On the other hand, a nozzle with fewer points may look dull.

(01-1) Piping rosettes

Rosette is also called 'piping roses.' Sometimes a cake is covered with cream piped in rose shapes, and this is called 'rosette' or 'piping rosettes.' 'Piping rosettes' refers to a cake in which the same pattern is repeated. It can still be classified as a type of patterned cake because it gives a different ambiance depending on the direction of viewing or the way the cream is piped. Cover the entire cake with white cream, or pipe in two colors of pink and white, or light mint and white to make a lovely cake.

Closed star nozzle D6K   **Density 85%**   **Angle of surface and nozzle** 90°   **Distance between surface and nozzle** 1 cm

1 Start at 1 cm between the surface and nozzle, and 90 degrees angle.

2 When it comes to piping whipped cream, the volume is important. When the cream touches the surface at the height of 1 cm, it will create a base.

3 Pipe as if to draw number 1 to make a base.

4 Rotate clockwise, continuing to pipe towards the 9 o'clock direction.

5 After piping till 9 o'clock location, finish by lightly touching the cream at the position where it started.

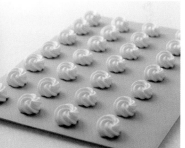

6 It is important to rotate in a small radius so there is no hole in the center when viewed from above.

(01-2) Piping traditional shells

{ The key to piping the traditional shell is to make a bulging head. Be careful not to open the angle too much because it can end up in a triangle shape. It is vital to pipe in uniform length, the head is bulging, and the tail at the end is sharp. }

Closed star nozzle D6K **Density** **85%** **Angle of surface and nozzle** 30° **Distance between surface and nozzle** 1 cm

1 Start at 1 cm between the surface and nozzle, and 30 degrees angle.

2 Flicking the wrist, curl the nozzle in a circle as if making a head-shape in the front.

3 When the round head is made, let the nozzle touch the surface.

4 Pipe softly to make a tail.

point ❶ When the board is lifted and viewed from the front, the length of the cream should be uniform, the head should be bulging, and should have a pointy tail at the end.

❷ If the cream is piped at a high angle, it may look like a triangle when viewed from above.

(01-3) Piping modified traditional shells

This is a technique to pipe a cute and chubby shell that complements the traditional shell and is more accessible. Squeeze the cream to fill the head and pull it out without applying pressure to finish it in a shape that resembles a balloon.

Closed star nozzle D6K | **Density** **85%** | **Angle of surface and nozzle** **30°** | **Distance between surface and nozzle** 0.5 cm

1 Start at 0.5 cm between the surface and nozzle, and 30 degrees angle.

2 Squeeze the cream without moving both hands.

3 Pipe the cream until it is completely round on the surface, then gently pull it out.

point The cream should look like a balloon when viewed from above. As with the traditional shell, if it is piped with an open angle, it may become a sharp triangular shape rather than a round balloon.

(01-4) Piping laid shells

{ Starting from the same position as '01-2. Piping traditional shells', pipe the modified shell with the nozzle turned to 5 o'clock direction. Start from half the height of the previously piped cream, pipe in one connected row. }

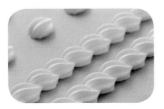 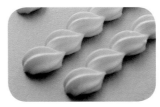

Closed star nozzle D6K   Density **85%**   Angle of surface and nozzle **30°**   Distance between surface and nozzle **0.5 cm**

1 From the position of '01-2. Piping traditional shells', turn the nozzle to 5 o'clock direction and start at 0.5 cm height.

2 Squeeze the cream without moving both hands.

3 Pipe the cream until it is completely round on the surface, then gently pull it out.

point It should keep a straight line when looked at from above. Pipe the cream continuously at half the height of previously piped cream to maintain a straight line.

⑴1-5 Piping hearts

It is a technique to make a heart-shape by modifying the traditional shell piping. It is also called 'point decoration' or 'one-sided decoration' because of its braided hair and knit-like shape. It is mainly used to give an accent in a single line, and for this type of decoration, it is better to pipe the cream at 1/3 of the way in on top of the cake. The key is to create a gap by piping the right side of the starting line lower than the left. Pipe continuously between gaps to avoid any empty space.

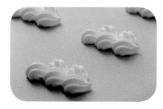

Closed star nozzle D6K   **Density** **85%**   **Angle of surface and nozzle** 30°   **Distance between surface and nozzle** ⊥ 1 cm

1 From the position of '01-2. Piping traditional shells', turn the nozzle to 5 o'clock direction to start.

2 Flicking the wrist, curl the nozzle in a circle as if making a head-shape in the front.

3 Pull softly to make a tail at the end.

4 Rotate the wrist in the opposite direction and find the second starting point to start piping.

5 Start the second piping just above the end of the first piped cream.

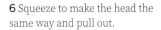

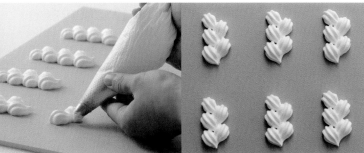

6 Squeeze to make the head the same way and pull out.

7 Rotate the wrist in the opposite direction again and find the third starting point between the gap of the first and second cream.

point Until the skill is adapted, move the body directly to change directions, and when you become proficient, move only the wrist to pipe in the opposite direction.

8 Repeat to finish piping.

point When viewed from above, the second cream should be lower than the first. This will open the gap for the third cream to start.

(01-6) Piping ropes

It is usually used to decorate with grass nozzles in the fall season. Keep the angle between the surface and the nozzle at 30 degrees, and spin as you pipe in one direction. It should be piped, not too big, or too tight to make a pretty shape. The cream must be curled well and not spread out when viewed from the opposite side.

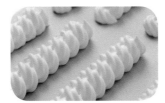
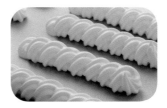

Closed star nozzle D6K **Density 85%** **Angle of surface and nozzle** **30°** **Distance between surface and nozzle** 1 cm

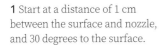

1 Start at a distance of 1 cm between the surface and nozzle, and 30 degrees to the surface.

2 Evenly distribute the strength of your hands and pipe the cream while rotating from the inside towards the outside.

3 After spinning and piping the cream, pull the cream out from the top to finish.

point ❶ When rotating the cream outward, it should not be turned too big so that it does not flare out and remain in a curled form.

❷ Check from the opposite direction of the piped cream, and make sure it is curled without flaring when viewed from the other side.

(01-7) Piping reverse shells

{ Reverse shell piping, looking like a number 8 turned sideways, is also called 'butterfly piping' or 'ribbon piping,' and it is a modified technique originating from 'rosettes.' Starting from the rosette piping, the key is not to rotate the cream until the end but pull it out and rotate in the opposite direction to make it overlap. }

Closed star nozzle D6K   **Density** 85%   **Angle of surface and nozzle** 90°   **Distance between surface and nozzle** 1 cm

1 Start at 1 cm between the surface and nozzle, and 90 degrees angle.

2 Pipe as if to draw number 1 to make a base, then turn clockwise, continuing to pipe.

3 Don't make a full turn, but pull it out at the 5 o'clock location.

4 Start the second piping at the position where the first piping ended and rotate counterclockwise to the 3 o'clock direction.

5 Pipe all the way to the 3 o'clock location, gently touch the cream, and pull out slowly.

<u>point</u> Same as the rosette, be careful not to rotate big so that hole is made in the center. The tail of the first piped cream should not be seen.

6 The first piped cream and the second piped cream should be the same size when viewed from above.

⟨01-8⟩ Piping connected reverse shells

{ This method is piping '01-7. Piping reverse shells' in a single line. It is a design that is loved again as custom cakes are back in trend. }

| Closed star nozzle D6K | Density 85% | Angle of surface and nozzle 90° | Distance between surface and nozzle 1 cm |

1 Start at 1 cm between the surface and nozzle, and 90 degrees angle.

2 Pipe as if to draw number 1 to make a base.

3 Turn clockwise, continuing to pipe.

4 Don't make a full turn, but pull it out at the 5 o'clock location.

5 Start the second piping at the position where the first piping ended and rotate counterclockwise to the 3 o'clock direction.

6 Pipe all the way to the 3 o'clock location, finish by gently touching the cream.

7 Start piping the third cream by slightly rising from the position where the second cream finished.

8 Now pipe while turning clockwise.

9 Pipe until 5 o'clock location, finish by gently touching the cream.

10 Pipe while turning counterclockwise again and repeat to finish.

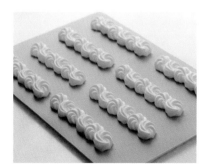

11 Turn the plate with piped creams at 90 degrees and observe. When the shells are piped in a connected form, it is not done right if the tails are visible. It is appropriately done when it looks like a single line.

(01-9) Piping modified connected reverse shells

{ Different impressions can be given just by changing the size of '01-7. Piping reverse shells.' Pipe the first cream small, the second cream larger, and repeat to complete piping. }

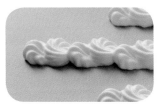

Closed star nozzle D6K | Density 85% | Angle of surface and nozzle 90° | Distance between surface and nozzle 1 cm

1 Start at 1 cm between the surface and nozzle, and 90 degrees angle.

2 Pipe as if to draw number 1 to make a base.

3 Turn small clockwise, as if to pipe a shell.

4 Don't make a full turn, but pull it out at the 5 o'clock location

5 At the end of the first cream, turn the nozzle bigger counterclockwise. Be careful not to create a hole in the middle.

6 Pipe all the way to the 3 o'clock location, finish by gently touching the cream.

7 Start piping the third cream by slightly rising from the position where the second cream finished.

8 Again, turn clockwise in small size and finish at 5 o'clock.

9 Pipe counterclockwise again, pulling it out at 3 o'clock location, and repeat to complete.

10 The finished cream should clearly show the difference between big and small. Be careful that the tails don't show.

02

Round nozzle

Round nozzles are most popular among all the nozzles and are always loved because these don't go out of fashion. In contrast to the easy piping, it can produce a significant effect and is often used as a lovely and cute decoration. If the creams are piped too small, it may look crowded, so piping a little larger can make it look more elegant. There is a lot to be aware of because the round nozzles do not have any points the cream comes out as it is piped.

(02-1) Piping teardrops

{ Teardrop piping is also called 'Kisses piping' because of its shape. It is a lovely and cute design. The key is to loosen the strength on the right hand and pipe it quickly to make a pointy horn. }

Round nozzle 809 **Density** **80~85%** **Angle of surface and nozzle** 90° **Distance between surface and nozzle** 1 cm

  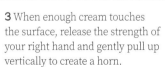

1 Start at 1 cm between the surface and nozzle, and 90 degrees angle.

2 With both hands still, pipe the cream so that it falls in a circle on the surface.

point When piping, lines can appear if hands are moving; therefore, without moving both hands, as soon as the cream touches the surface, lift up the nozzle immediately.

3 When enough cream touches the surface, release the strength of your right hand and gently pull up vertically to create a horn.

point The height and size should be uniform when viewed from the front. Be careful not to pipe too high or too low, and finish so that it looks like the roof of a palace.

⟨02-2⟩ Piping laid teardrops

{ Pipe the same way the laid shells are piped with the closed star nozzles. After piping the cream into a plump and perfectly round shape, relax your hand and pull out to make a horn. This is the most used technique to decorate the border on a round cake. }

Round nozzle 809   **Density** 80~85%   **Angle of surface and nozzle** 45°   **Distance between surface and nozzle** 0.5 cm

1 Starting from '01-2. Piping traditional shells' position, pull the nozzle toward your body at 5 o'clock and start at 0.5 cm up from the surface.

2 Pipe the cream without moving both hands.

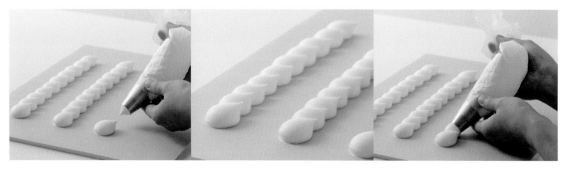

3 Pipe the cream until it's fully round and pull it out lightly.

4 The cream's head should not float, but touching the surface.

5 Start the second piping at half the height of the first cream.

6 In the same manner, when the cream is fully round on the surface, gently pull it out.

7 Start the third piping at half the height of the second cream.

<u>point</u> When piping laid shells with a closed star nozzle, the angle between the nozzle and the surface is 30 degrees. But when using a large round nozzle, the angle must be opened more to about 45 degrees. That way, the cream can reach the surface in a short and plump shape.

8 Using the same method, when the cream touches the surface, pull it out lightly.

<u>point</u> If the piping elongates, it will not look like a teardrop. The piping is properly completed when it looks like round water drops have fallen on a surface.

03

French star nozzle

Along with the round nozzle, it is a widely used nozzle. It's relatively easy to handle due to its multiple points. If the round nozzle gives a simple and cute impression, the French star nozzle gives a gorgeous and cute impression. The French star nozzle is also often used to pipe teardrops. By changing the size of the nozzles depending on the type of decoration, you can make a cake with a high degree of completion.

(03-1) Piping teardrops

{ The French star nozzle is used in a similar way to the round nozzle, but it is easier to work with compared to the round nozzle. Be careful that if the right-hand shakes or moves while piping, lines will appear, and if the nozzle is pulled out slowly, not quickly, the tip of the horn will come out blunt. }

French star nozzle 867K **Density** **85%** **Angle of surface and nozzle** 90° **Distance between surface and nozzle** 1 cm

1 Start at 1 cm between the surface and nozzle, and 90 degrees angle.

2 With both hands still, pipe the cream so that it falls in a circle on the surface.

3 When enough cream touches the surface, release the strength of your right hand and gently pull up vertically to create a horn.

point Note that if the cream is piped at a low height, it will become stumpy, and if piped from a too high position, the cream will elongate and bend.

⑩3-2 Piping laid teardrops

{ It works the same way as piping laid teardrops with a round nozzle, but it is easier to make, and the finished shape is prettier. If the cream piped with an 869K nozzle seems big, you can also use an 867K nozzle, which will make a more neat shape. }

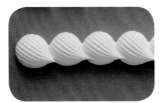 

French star nozzle 867K   **Density** 85%   **Angle of surface and nozzle** 45°   **Distance between surface and nozzle** 0.5 cm

1 Starting from '01-2. Piping traditional shells' position, pull the nozzle toward your body at 5 o'clock and start at 0.5 cm up from the surface. Pipe the cream without moving both hands.

2 Pipe the cream until it's fully round and pull it out lightly.

3 Start the second piping at half the height of the first cream.

4 When the cream is piped fully round, pull it out lightly.

5 Pipe the third cream the same way.

6 Likewise, when the cream is piped fully round, pull it out lightly.

7 Complete the same way, taking care that the head of the cream in the starting position of the piping does not float from the surface.

(03-3) Piping hearts

{ Done in the same way as piping hearts with a closed star nozzle, it mainly gives an accent in one row on the cake. Because it is used for accented decoration, it is better to pipe with 869K rather than 867K for a plump look. It can be piped in a connected row or pipe in three to make a cute shape. }

French star nozzle 867K **Density 85%** **Angle of surface and nozzle** **45°** **Distance between surface and nozzle** **1 cm**

1 Start at a distance of 1 cm between the surface and nozzle, maintaining 45 degrees angle, pull the nozzle to the 5 o'clock position.

2 Flicking the wrist, curl the nozzle in a circle as if making a head-shape in the front.

3 Pull softly to make a tail at the end.

4 Turn the wrist in the opposite direction and locate the starting point.

5 Start the second piping at half the height of the first cream.

6 In the same way, curl as if making a round and bulging head.

point Until the skill is adapted, move the body directly to change directions, and when you become proficient, move only the wrist to pipe in the opposite direction.

7 Start piping the third cream between the gap of the first and second cream.

8 Start piping the same way as the first to finish piping.

04

Ruffle nozzle

Ruffle nozzles are compelling for designs that give an elegant and fancy impression.
There are various nozzles available, such as numbers 112, 119, 123, 104, 96, 125K, etc.
Depending on the nozzle, it can make a cute or colorful ambiance.

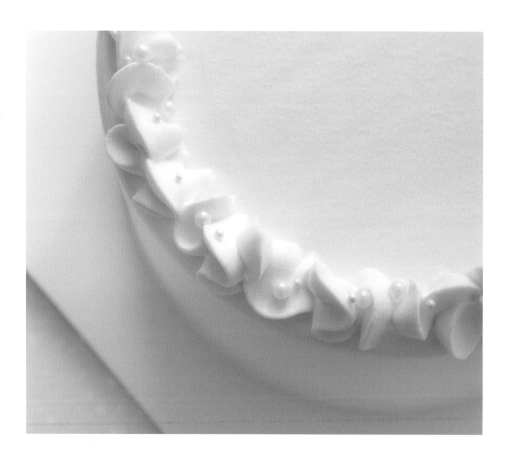

(04-1) Piping basic ruffles 1

{ This is the very basic of ruffles. Number 114 leaf nozzle is usually used, but the piped cream may look dull if it comes out too big. Number 112 nozzle is used for this book. It can also make a cute ribbon pattern. }

Leaf nozzle 112 Density 85~90% Angle of surface and nozzle 30° Distance between surface and nozzle 0.5 cm

1 Start at 0.5 cm between the surface and nozzle, and 30 degrees angle.

2 Pipe the cream from left to right.

point The waved shapes should look uniform when viewed from above. Be careful not to pipe too loose. It is also critical to give even strength in your hand.

(04-2) Piping basic ruffles 2

{ It is a technique of piping out evenly in a single line. It can give a lot of shading effects, so even one line can look gorgeous. }

Petal nozzle 125k, 104   **Density** 85~90%   **Angle of surface and nozzle** 30°   **Distance between surface and nozzle** 1 cm

1 Start at 1 cm between the surface and nozzle, and 30 degrees angle.

2 Pipe the cream from left to right.

point ❶ The waved shape should look uniform when seen from the opposite side. Keep the strength of your hand evenly.

❷ To shape the cream beautifully by the cream flowing out of the nozzle, make sure not to pipe the cream too low.

(04-3) Piping small ruffled flowers

{ Usually, nozzle numbers 102, 103, and 104 are used for small ruffles. It could bloom neat, round, and cute petals. There are three types of Wilton 104 nozzle; Wilton, Wilton Korea, and Wilton China. Particularly, the Wilton China nozzle works well for expressing plump and cute ruffles. }

Petal nozzle 104 | Density 85~90% | **Angle of surface and nozzle** 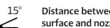 15° | **Distance between surface and nozzle** 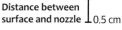 0.5 cm

1 Hold the narrow part of the nozzle upward, keeping 0.5 cm distance between the surface and nozzle, and 15 degrees to the surface, pipe the cream while moving up and down.

2 It should feel like the cream is spewing out automatically, not myself piping it out.

3 Pull out so that it can be finished with a sharp end.

point ❶ When starting to pipe, be cautious not to widen the angle with the surface because it can make the end round.

❷ While keeping the center point without moving your body, make a flicking movement with your wrist to let the cream pipe out up and down to create natural ruffles.

❸ If you start piping from too high, too much cream will pipe out, looking like a flat noodle. Make sure to check when the piped cream is viewed from the opposite side that it looks like blooming round petals.

(04-4) Piping connected small ruffled flowers

{ Connected small ruffles flowers can be styled as a wreath-shaped decoration around the edge of the cake or filling the top of the cake completely. }

Petal nozzle 104 **Density** 85~90% **Angle of surface and nozzle** 15°   **Distance between surface and nozzle** 0.5 cm

1 Hold the narrow part of the nozzle upward, the bottom end almost touching the surface.

2 Pipe the cream adjusting the size and height to make fancy ruffles.

3 When starting to pipe, be careful not to open the angle of the nozzle.

<u>point</u> When viewed from the other side, you should be able to see natural small petaled ruffles. Make sure not to squeeze out the cream from too high.

(04-5) Piping barley ears

{ It's a good design to make it into a wreath-shape that neatly wraps around the edge of the cake. Use nozzle number 104 or 103. }

Petal nozzle 104   **Density** **85%**   **Angle of surface and nozzle** 15°   **Distance between surface and nozzle** 0.5 cm

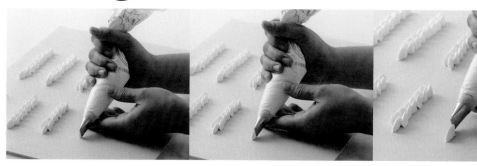

1 Hold the narrow part of the nozzle upward, the bottom end almost touching the surface, and keeping the nozzle at 15 degrees to the surface.

2 Hold the nozzle sideways and pipe crosswise, applying the same way as laid piping.

3 Pipe the cream not too big, so it looks neat and connects in one line.

point ❶ It looks simple and prettier to complete with small nozzles rather than larger ones.

❷ It is vital to cross pipe to avoid gaps.

(04-6) Piping waves

{ It is a shape of overlapped rose petals and is usually used as wreath-style decoration. This design can give a neat and cute vibe. }

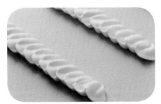

Petal nozzle 104 　 Density **85%** 　 Angle of surface and nozzle **15°** 　 Distance between surface and nozzle **0.5 cm**

1 Hold the narrow part of the nozzle upward, keeping 0.5 cm distance between the surface and nozzle, and 15 degrees angle, start piping while turning the nozzle clockwise.

2 After blooming one petal, immediately pipe the second petal overlapping to the first one.

3 Pipe by connecting the petals, but not clumped together so that it looks like petals when seen from above.

4 Pipe neatly to make all cream looks like petals.

(04-7) Piping ruffled waves

> Using a leaf nozzle number 95 lets you create rippling wave shapes. This nozzle is usually used for cupcakes, but it can be held differently and used for round cakes or square cakes to create a stylish design.

Leaf nozzle 95 **Density** 85~90% **Angle of surface and nozzle** 15° **Distance between surface and nozzle** 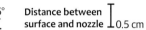 0.5 cm

1 Hold the nozzle vertically, keeping 0.5 cm distance between the surface and nozzle, and 15 degrees angle.

2 Flick the wrist back and forth, pipe quickly from left to right.

3 It looks fancier and prettier piping in irregular size than the uniformed size.

point **❶** Use the clustered side of the cream as the backside when used for decoration.

❷ If the density of the cream is too thin, the creases will look blurred. The density is quite thick amongst the piping cream.

05

Pattern making nozzle

Pattern making nozzles can be used to decorate the top of the basic round cakes,
but in Congmom's Cake Design class, it is mainly used on the square cakes.
This chapter will cover line, curved pattern, and embossing pattern,
cute ribbons, heart pattern, small flower, and big ruffle patterns.

(05-1) Piping straight and curved lines

{ Small round nozzles are used, such as numbers 0, 00, and 1. It is a method of drawing straight, and curved lines with cream piped 1 cm above the surface. }

Round nozzle 00	Density 80~85%	Angle of surface and nozzle 90°	Distance between surface and nozzle 1 cm

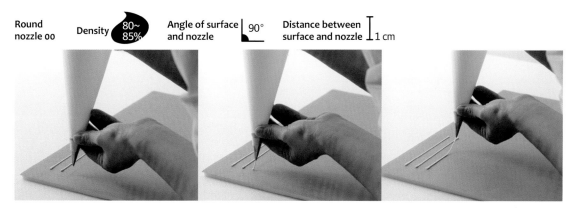

1 Start at 1 cm between the surface and nozzle, and 90 degrees angle.

2 Maintain the grip evenly, so that the cream does not break, and pipe so that a line can be formed as the cream flows from the nozzle.

3 Finish by lowering the nozzle so it touches the surface.

4 When piping curved lines, start from 1 cm above the surface, draw freely, and then lower the nozzle to finish.

(05-2) Piping big ribbons

In the case of ruffles, the cream is usually spouted from left to right, but it is natural to pipe vertically rather than horizontally for the large ribbons. For these, 125K nozzles are mainly used. The opening of the nozzle is larger than the others, so if the density of the cream is thin, it tends to sag. It is stable to use a thicker density of cream than the small ruffles.

Petal nozzle 125K   Density 85~90%   Angle of surface and nozzle 15°   Distance between surface and nozzle 0.5 cm

1 Hold the narrow part of the nozzle upward, keeping 0.5 cm distance between the surface and nozzle, and 15 degrees angle.

2 Unlike the previous methods, squeeze the bag harder as if the cream is pouring out.

3 Finish the ribbons as if the flowing cream is naturally piping by itself, not myself shaping the cream.

point ❶ Check the piped cream from the opposite side. It should have fancy lines when seen from the opposite side.

❷ Make sure the cream does not sag. Rather than piping it evenly, try to make a large natural ribbon.

(05-3) Piping ribbons

{ You can make ribbons with nozzle number 112. It is good to use as decoration when you want to add an accent to the connecting part of the ruffle on a customized cake. The lines are clearly visible when viewed from the opposite direction. }

 Leaf nozzle 112 **Density** 85~90% **Angle of surface and nozzle** 15° **Distance between surface and nozzle** 0.5 cm or less

1 Hold the nozzle almost touching the surface, angled 15 degrees to the surface.

2 Pipe the first ribbon big to form a natural wing shape.

3 Pipe the second cream continuously to overlap the first cream.

4 Pipe the third ribbon, pull the nozzle softly to finish.

point ❶ Piping a total of three times, stacking while controlling the cream, so the amount gradually decreases.

❷ If the density is too thin, the ends will stick together without splitting. The density of the cream should be thicker than the icing cream so that the ends naturally split to form a ribbon.

⑤-4 Piping hearts 1

{ Hearts can be made with a small leaf nozzle. The size of the nozzle doesn't matter; you can use any size you want or any small size nozzle you have. If the nozzle is not available, it can be replaced with a piping bag or cornet by cutting off the tip. }

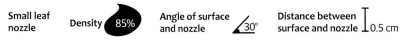

| Small leaf nozzle | Density 85% | Angle of surface and nozzle 30° | Distance between surface and nozzle 0.5 cm |

1 Start from 0.5 cm between the surface and nozzle, and 30 degrees angle.

2 Without moving both hands, pipe the cream on the surface.

3 When it becomes a heart shape, pull out the nozzle softly to make a pointy end.

(05-5) Piping hearts 2

You can make standing hearts by piping the small leaf nozzle vertically, not laid to the side. Since it's a small pattern, it can be placed not only on shortcakes but also on tarts and petit mousse cakes. The size of the nozzle doesn't matter; you can use any size you want or any small size nozzle you have.

Small leaf nozzle | **Density** 85% | **Angle of surface and nozzle** 90° | **Distance between surface and nozzle** 0.5 cm

1 Start from a distance of 0.5 cm between the surface and nozzle, and 90 degrees to the surface.

2 Without moving both hands, pipe the cream until it touches the surface in a round shape.

3 When it becomes a heart shape, pull out the nozzle softly to make a pointy end.

<u>point</u> ❶ When seen from the front, the height and size should be uniform.

❷ If the cream is too thick, the pointy end will split into two, so make the cream is thinner than the one used for the ribbon piping.

06

V-shaped (St. Homoré) nozzle

I usually use numbers 481 and 581 are used for V-shaped (St. Honoré) nozzles, and because it makes dimply shapes, I call it 'embossing piping.' By holding the nozzle at 90 degrees from the surface, pipe it as the shell-piping method. A nozzle that the ridge of the V-shape is not too deep should be used, like no.581, to pipe plump creams. It can be used in three ways: piping in a regular pattern, piping line-by-line in a different direction, and piping irregularly.

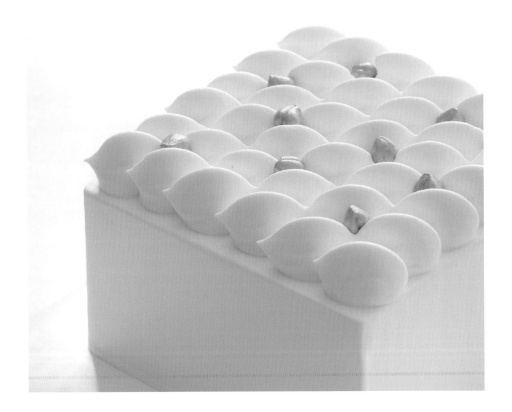

⑥⁶⁻¹ Piping basic embossing

{ Pipe in the typical method of voluminous style. The nozzle can be replaced by cutting the tip of a piping bag. }

V-shaped(St. Honoré)
nozzle 581 Density 80~85% Angle of surface and nozzle 90° Distance between surface and nozzle 1 cm

1 Start from 1 cm between the surface and nozzle, and 90 degrees angle.

2 Without moving both hands, pipe the cream until it touches the surface in a round shape.

3 When the cream becomes plump, pull out the nozzle softly to make a pointy end.

4 Pipe from the middle of the previously piped cream and finish by making a pointy end. Repeat.

⑥-② Piping weaved embossing

{ This technique is braiding the cream in different directions, line by line. It can give a fancier feel than piping the basic embossing. Braid so that there are no empty spaces. }

V-shaped(St. Honoré) nozzle 581　Density 80~85%　Angle of surface and nozzle 90°　Distance between surface and nozzle 1 cm

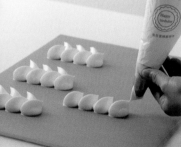

1 Pipe vertically applying the same technique as '06-1. Piping basic embossing.'

141p

2 Turn your body in the opposite direction and start over piping from the tail of the last cream.

3 Pipe one line the same way.

Tip

① Volume is vital when it comes to piping emboss pattern. Make sure not to pipe the cream too elongated, and the density of the cream is soft so that the tails at the end are sharp.

② Completing a uniformed pattern look neat and tidy, but it becomes an interesting pattern even if it's piped irregularly.

07

Drop flower nozzle

Use nozzle number 108 and pipe the same way as piping teardrops.
A cute dumpling shape can be made.

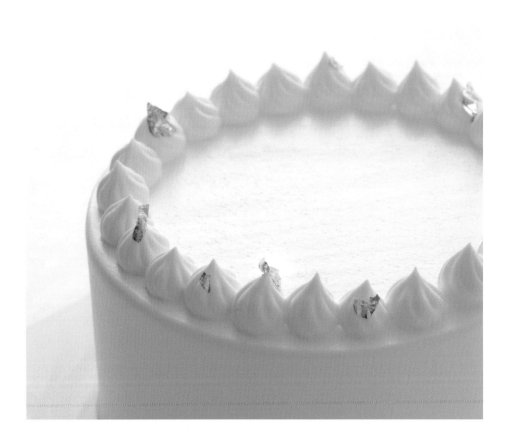

Piping drop flowers

{ Drop flower nozzles make neat and cute shapes, which can be decorated as a wreath on the border of the cake. Refine the shape while paying attention to the height and density. }

Drop flower nozzle 108   Density 80~85%   **Angle of surface and nozzle** 90°   **Distance between surface and nozzle** 0.5 cm

1 Start from 1 cm between the surface and nozzle, and 90 degrees angle.

2 Pipe the same way as '02-1. Piping teardrops.' **117p**

point Normally, teardrops are piped from 1 cm above the surface, but since nozzle number 108 is relatively small, and if the cream is piped at 1 cm height, it may come out too long. Therefore, it is necessary to pipe at a much lower height than the conventional method to make a plump and cute shape.

3 To finish, lightly lift the nozzle vertically to make a pointy horn.

Decoration Techniques
Using Other Tools

In addition to the various nozzles shown previously, the decoration becomes more interesting if other tools are used, such as a spatula and measuring spoons. The side of the cake can also be designed with a variety of patterns using a fork, brush, triangle, and rectangle cake combs.

01

Decoration using a spatula

For the scooping technique using a spatula, it is easier to scoop from a wide bowl rather than a deep one. For this method, the consistency of the cream is most important, which should be adjusted to the consistency of a strong yet soft cream that can hold the pointy horn. The scooping method is usually used on chiffon cakes. In the case of the chiffon cake, it is good to take advantage of the characteristic of having a hole in the middle. Congmom recommends petal-scooping.

Spatula   Skill ●●●●●

(01-1) Scooping peaches

{ After organizing the texture of the cream, use a spatula to scoop the surface of the cream twice in a row to create a tasty and cute peach shape on the cake. It can be used on tarts, petit mousse cakes, or plated desserts. }

1 In a wide bowl, gather the cream forward with a spatula and spread it evenly from side to side.

2 Hold the spatula vertically at 90 degrees and scoop the cream off the surface twice in a row using only the front part of the spatula. The second cream should have more cream than the first in order to make a clear line in between.

3 Place the cream at 90 degrees on top of the cake and let the bottom of the cream completely sit on the cake.

4 When the cream touches the surface, gently lift the spatula and remove it.

(01-2) Scooping dumplings

{ If you've become proficient at making peaches, now try making dumplings by scooping the cream four times in a row. }

1 In a wide bowl, gather the cream forward with a spatula and spread it evenly from side to side.

2 Hold the spatula vertically at 90 degrees and scoop the cream off the surface using only the front part of the spatula.

3 Start scooping the surface the same way as '01-1. Scooping peaches.' 149p

4 Scoop the surface four times in a row. The amount of cream being scooped is equal.

5 When the cream touches the surface, gently shake it off and softly lift the spatula to remove.

When proceeding '01-1. Scooping peaches' and '01-2. Scooping dumplings', please refer to the following.

① When placing the cream on the cake, it should be the cream that is vertical to the cake, not the spatula. The cream is perfectly set when the spatula is slightly tilted to the left, and the cream is vertical to the cake.
② The cream should be scooped on the tip of the spatula to make lovely peaches and dumplings.

(01-3) Scooping petals 1

{ Petal shape scooping gives a neat and cute impression. It may look simple and a little flat, but a neat flower can be made into a decoration by using the hole in the chiffon. }

1 In a wide bowl, gather the cream forward with a spatula and spread it evenly from side to side.

2 Hold the spatula flat; only the front blade of the spatula will be used.

3 Stick the spatula into the cream diagonally.

4 Pull out as is, diagonally as it was put in.

point The spatula must be pulled out diagonally, not vertically, to make a nice horn.

5 When the cream is on the blade, turn the spatula over.

6 Place the pointy horn slightly on the edge of the cake, gently press the tip of the spatula with your left hand to spread out the cream.

point You should control the amount of cream well so that the petal doesn't flare out. If too little cream is scooped, it will not reach the hole of the cake, and if too much cream is used, the tip of the petal will follow the spatula.

7 Pull out lightly to finish a natural petal.

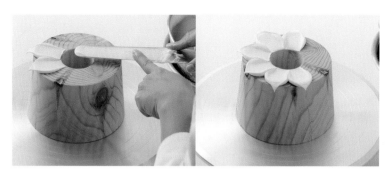

8 Let the second petal naturally overlap with the first petal.

9 Rotate a full turn to complete.

point For a 15 cm chiffon cake, eight petals look the best.

(01-4) Scooping petals 2

{ It is a bold petal making method that can maximize the shading effect. Scoop the cream by holding the spatula upright at 90 degrees, then scrape the back edge of the spatula off the bowl while pulling out will create a horn at the tip of the cream. Place the spatula at 90 degrees on the cake and tilt it slightly to the side, apply pressure to the front of the spatula and pull out to make a glamorous petals. }

1 In a wide bowl, gather the cream forward with a spatula and spread it evenly from side to side.

2 Hold the spatula upright at 90 degrees, scoop the cream with the blade and pull it out straight.

point At this time, scrape the spatula off the bowl to form a natural horn.

3 The horn is made.

point The horn shouldn't be too long. If the cream is too thin, the horn will bend backward; too thin, it won't appear.

4 Place the cream vertically on the cake.

point When the cream and cake are vertically placed, the angle of the spatula opens naturally.

5 Let down the spatula so the cream touches the cake.

6 Tilt the spatula to the left to push the cream, and pull it out while taking care not to let the cream come with the spatula.

7 Make the second petal the same way.

8 For a 15 cm chiffon cake, finishing with 8~9 petals look the best.

point If the petals are placed too tight or make too many petals, the cake may look stuffy or feel excessive.

02

Decoration using a fork

For decoration that uses a fork, a lightweight fork made of wood or plastic is more suitable than a heavy fork made of stainless steel. Because when decorating whipped cream cakes, if the tool is too heavy, it takes more strength to hold, causing it to dig in the light-textured cream. Also, shorter forks are more accessible to control than longer ones.
You can use the fork to make lines on the bottom or middle of the cake, and in the winter season, you can make a log-shaped Buche de Noel cake in white color.

Fork   Skill ●●○○○

1 Prepare a fork with four points.

2 Place the back of the fork on the bottom of the cake.

3 Relax your hand and rotate the turntable so that the fork lightly draws the lines.

4 Draw lines in the middle as well while being careful not to shake your hand.

point If your hand is shaking while rotating the turntable, the lines drawn on the cake also shakes, and the finished cake will not look pretty. To prevent your hand from shaking, attach your arm to your torso.

5 You can draw freely on the side of the cake to make Buche de Noel cakes for the winter.

03

Decoration using a measuring spoon

The measuring spoon can be heated with a blowtorch or warmed in hot water and then wiped dry to make grooves on the iced whipped cream. The technique is to melt the cream just enough to form the shape, so if the temperature of the spoon or measuring spoon is too high, the cream will melt too much and go out of shape. For beginners, it is recommended to dip the tool in hot water rather than using a blowtorch.

Measuring spoon  **Skill** ●●●●●

1 Prepare hot water, measuring spoon, and dry cloth.

2 Pipe teardrops with a round nozzle.

3 Keep the measuring spoon in hot water until hot enough, then wipe dry with a cloth.

4 Hold the spoon on top of the cream and press lightly, and melt the cream.

5 Slightly turn and lift the spoon to make a groove.

<u>point</u> You can fill the groove with ganache or jelly to make a fancy decoration.

04

Decoration using a plastic card

You can make a rectangular card by cutting a cake collar or stiff plastic file.
Working the same way as decorating sides using a scraper can give cake comb effects to
dome cakes, tall round cakes, and tree-shaped cakes. The key is to make delicate lines.
Keep checking the cream gathering on the tip of the plastic card,
and practice while correcting the wrong posture.

Plastic card  Skill ●●●●○

1 Prepare a stiff plastic card, not too long, and place it diagonally on the bottom side of the cake.

2 With the card tilted diagonally, rotate the turntable and move the card up from the bottom to the top.

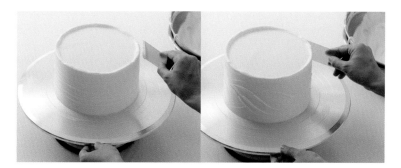

3 Relax your hand while rotating the turntable, so the lines are created naturally on the side of the cake.

point Make sure to keep your hand still while turning the turntable. If you give too much pressure or open the angle of the card outward, the cream may get dug in deep, and the cake sponge may become visible. The strength to hold the card is enough.

4 You can make wave patterns by moving the card up and down while rotating the turntable fast.

Ø5

Decoration using a brush

When decorating with brushes, it is better to use short and lightweight brushes. The length of the bristles, about 36 mm, works well, and it is more convenient to decorate with a wide brush rather than a narrow brush. Moving the brush up diagonally makes grain patterns. You can give impressions of wings and feathers; also, you can make it more colorful by dipping the brush in a colored whipped cream.

Brush   Skill ●●○○○

1 Hold the brush lightly as if holding a pencil. Use only three fingers- thumb, index finger, and middle finger- not to give too much pressure.

2 Start from the facing-side front. From left to right, move diagonally to make a grain pattern.

<u>point</u> When moving the brush up, make sure to brush in a straight line. Be careful not to brush in a curved line.

3 After completing the side, trim the crown on top.

<u>point</u> The brush marks appear clear when they are made immediately after icing the cake. If the brush is used after some time has passed, the dried cream may stick and cause it to get messy.

06

Decorations using rectangle and triangle cake combs

By using a cake comb or triangle comb, you can give various effects to the side of the cake. You can even make your own design. Cake combs are more effective in decorating sides of a tall cake or a chiffon cake than a low-height cake. It is because if you use it on a low-height cake, it makes it look flat overall. When emphasizing the side of the cake, it is better to decorate the top as simple as possible.

Cake comb, Triangle comb  | **Skill** ●●●○○

1 Prepare a light plastic triangle comb. **2** Place the comb on the side of the cake, opening the angle as little as possible, and rotate the turntable to start icing. **3** Pull out in a fast motion to finish.

4 Try using various cake combs as well.

point ❶ If the design on the comb is asymmetric, you can choose the direction on the cake, up or down, to ice.

❷ The cream needs to be dug deep to emphasize the design, in which a thicker layer of cream should be applied when icing the sides than the standard method.

LESSON 08

Congmom's Cake Diary

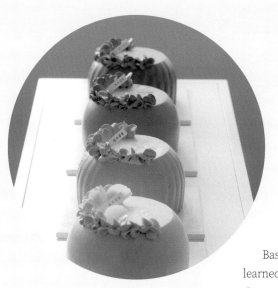

Based on the techniques
learned earlier, let me introduce
Congmom's Cake Design, which
combines various decorations that go
well with the design. The know-how of
Congmom's Cake Design class has been
transferred as is, with 33 designs and
four application methods
loved by many.

01

Basic cake finished with laid piping technique

Tool used	**Technique**	Laid piping	**Piping angle**	45 degrees
French star 867K	**Density**	85%	**Skill**	●●●○○

ingredients & tool

peaches, oregano leaves, silver leaves, fruit-baller, brush

design recipe

1   Prepare a 15 cm round cake, icing completed. `54p`
2   Start from 0.5 cm above the top of the cake, held at 45 degrees. `122p`
3   Pipe the cream without moving both hands.

1   2   3

Congmom's Cake Diary • 169

design recipe

4   Pipe the cream until it is completely round on the surface, then relax your hand and gently pull it out.

5   Start piping the second cream over the tail of the first cream.

6   While turning counterclockwise, pipe on the edge of the cake.

7   Scoop out a peach in round shape. Use a 22 mm diameter fruit-baller to give a cute and tidy look, which goes well with a 15 cm cake.

8   Press the peach firmly with the fruit-baller. Press firmly until the juice leaks through the hole in the middle of the fruit-baller.

9   Twist the fruit-baller and dig out the peach into rounds.

tip

⑤ Make sure that the tail of the piped cream is not extended towards the edge of the cake. It's good to roll the ends to give it a soft look.

⑥ Pipe 1~2 mm inside from the edge of the cake. Piping too inward can make the finished top look stuffy and tight; conversely, piping too close to the edge or go over the edge line may cause the piped creams to collapse or melt and flow down.

4   5   6

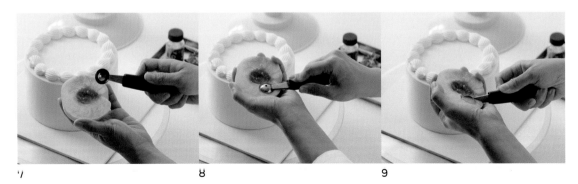

7   8   9

design recipe

10 The peach for decoration is scooped round.

11 Place the scooped peaches on a kitchen towel and use it after removing moisture.

12 When decorating, place the cake on the turntable and start from the front side after deciding where you want it to be.

13 This cake will be finished with a point decoration (one-sided) that is biased only to one side to give a tidy look. Place the peaches on the top left.

14 Stack the peaches to look voluminous.

15 Top with oregano leaves to give it freshness.

tip

⑮ Herbs look prettier to use the backside. The color is more subdued than the front side, and the leaf veins are lovely, which goes well with the white cream.

10   11   12

13   14   15

design recipe

16 Finish with few silver leaves on the peach to give it a cool effect.

17 Place a cake topper where the decoration is so that the attention can be directed towards one place. A round topper will go well with the round peaches.

tip

⓯ Use gold or silver leaves just enough to give points. If too much is used, it can look rather messy.

16   17

02

Basic cake finished with teardrop piping technique

Tool used
French star 869K,
Round 809

Technique | Teardrops		**Piping angle** | 90 degrees	
Density | 80~85%		**Skill** | ●●●○○	

ingredients & tool

figs, thyme, gold leaves, brush

design recipe

1   Prepare a 15 cm round cake, icing completed. `54p`

2   First, fill a piping bag fitted with the round nozzle. Start from 1 cm above from the top of the cake, held at 90 degrees. `117p`

3   Pipe out the cream without moving both hands while removing it vertically.

1   2   3

design recipe

4   Pipe 1~2 mm inside from the edge of the cake to prevent the cream from falling and looks stable.

5   Making sure not to pipe the cream too far inside, set guides as you pipe with the round nozzle.

6   Now fill a piping bag with cream, fitted with the French star nozzle.

7   In the empty space left after piping with the round nozzle, fill in irregularly using the French star nozzle. **121p**

8   Divide the figs into eight pieces and place them on the left side of the cake to decorate.

9   In between the figs, place a few long and thin stems of thyme to give a point.

tip

❼ For decoration that uses a round nozzle and French star nozzle together, it is best to set the guide with a round nozzle that is difficult to handle, then pipe with an easier-to-handle French star nozzle to finish. If the French star nozzle is used first, it's not easy to work with the round nozzle.

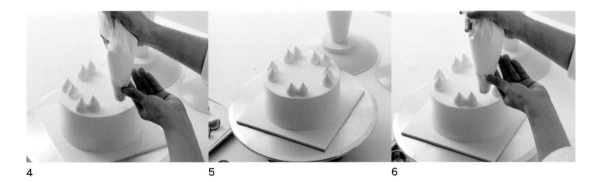

4   5   6

7   8   9

design recipe

10 Place some gold leaf on the figs, which goes well with the figs and gives a luxurious feel.

11 Place a cake topper where the decoration is so that the attention can be directed towards one place. A neat rectangular topper goes well.

tip

❿ Use the silver leaf to give a playful and cool note or gold leaf to provide a warm and luxurious note.

10   11

03

Basic cake finished with single lime heart piping technique

Tool used French star 869K	**Technique** | Piping hearts in a single line (applying traditional shell piping) **Piping angle** | 45 degrees
	Density | 85% **Skill** | ●●●○○

ingredients & tool

green grapes, oregano leaves, silver leaves, brush

design recipe

1 Prepare a 15 cm round cake, icing completed. **54p**

2 Fill a piping bag fitted with the French star nozzle. Start from 1 cm above the top of the cake, held at 45 degrees.

3 Start piping hearts, applying traditional shell piping. **124p**

1   2   3

완두콩맘

design recipe

4   Start piping as if to weave, alternating from left to right. Be careful not to make an empty space at this stage.

5   For a 15 cm cake, it looks stable to pipe seven times, starting from the left side and ending on the left.

6   Randomly place green grapes and oregano leaves.

7   Top with few silver leaves that goes well with the grapes and gives a cool effect.

8   Place a cake topper where the decoration is so that the attention can be directed towards one place. A round topper will go well with the round green grapes.

tip

❹ Pipe the front part of the cream while curling as if making a head-shape, then pull out to make a tail.

❼ It looks pretty to use green grapes cut in half as decoration because the inner flesh looks beautifully transparent.

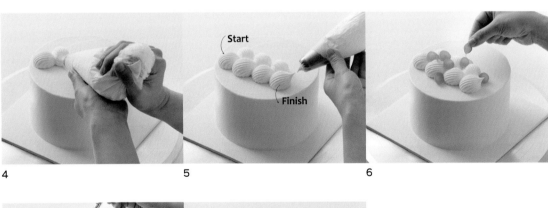

4   5   6

7   8

04

Basic cake finished with Closed star nozzle

Tool used
Closed star
D6K

Technique	Modified connected reverse shells	Piping angle	90 degrees
Density | 85%		**Skill** | ●●●●○	

ingredients & tool

marigolds

design recipe

1   Prepare a 15 cm round cake, icing completed. `54p`

2   Start from 1 cm above the top of the cake, held at 90 degrees. `114p`

3   Pipe the first cream rotating the nozzle clockwise in a small size.

1   2   3

design recipe

4   Start the second cream from the point where the first piping ended, turn big counterclockwise.

5   While maintaining the piping angle, repeat piping small and large pairs by rotating the turntable.

6   Because the big and small sizes of the creams are significantly different, it looks much prettier and not monotonous.

7   Piping is finished.

8   Since the pattern itself gives a sufficiently fancy ambiance, I will finish it neatly. Prepare yellow marigold petals.

9   When decorating, it is better to decorate it facing the front part of the cake, instead of turning the cake to change the direction.

tip

❹ The starting point of the first piping is essential. It's good to start from where the cake topper will be placed. That way, the boundary between the starting and the finishing point can naturally end with the cake topper.

❺ When practicing on a low surface or flat plate, it is correct to set the angle to the surface at 90 degrees. However, on the round cake, the angle should be slightly lowered so you can see the shape of the finished cream from the side as it is tilted slightly rather than pointing to the sky.

❼ Finally, the key is that the tail of the piped cream is not visible and connected in a single line.

4　　　　　　　　　　5　　　　　　　　　　6

7　　　　　　　　　　8　　　　　　　　　　9

design recipe

10 It looks more natural to place the petals irregularly rather than regularly.

11 The cake topper should be placed at the point where the piping started and ended meet so that the boundary does not stand out and vfinish naturally.

tip

⓫ A clean rectangular topper looks best for a neat looking cake.

10   11

05

Basic cake finished with Drop flower nozzle

Tool used
Drop flower
108

Technique	Drop flower/ Dumpling	Piping angle	90 degrees
Density	80~85%	Skill	●●○○○

ingredients & tool

gold leaf sheets, brush

design recipe

1   Prepare a 15 cm round cake, icing completed. `54p`

2   Start from 0.5 cm above the top of the cake, held at 90 degrees.

3   Pipe out the cream without moving both hands and remove quickly to make a horn. `145p`

tip

❷ Normally, teardrops are piped from 1 cm above the surface, but nozzle number 108 is relatively small. Therefore, it should not be piped too high for a good result.

1   2   3

design recipe

4   The wreath type decoration looks stable when the cream is piped in about 1~2 mm from the outer edge of the cake.

5   Piping is complete. The size should be uniformed and should have pointy horns.

6   To give it a clean look, I will only decorate it with gold leaves.

7   Place gold leaves randomly. You can put it on top of the horn as well.

8   It looks clean and neat. The wreath-type decoration is simple, but it can be completed beautifully.

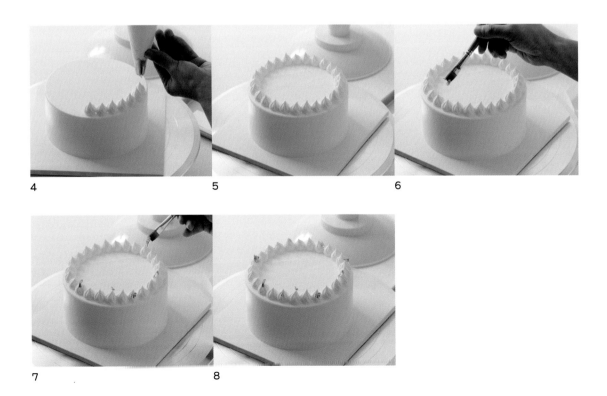

4   5   6

7   8

06

Basic cake finished with French star nozzle 1

Technique | Modified connected reverse shells		**Piping angle** | 90 degrees
Density | 85%		**Skill** | ●●●●○

Tool used
French star
867K

ingredients & tool

edible flowers, silver leaves, brush

design recipe

1 Prepare a 15 cm round cake, icing completed. `54p`

2 I will pipe modified connected reverse shells using an 867K nozzle. Start from 1 cm above the top of the cake, held at a 90-degree angle. `114p`

3 A total of 3 rows will be piped. Plan out the location to be completed, then pipe the middle row first.

tip

❷ For this cake, I will pipe in a horizontal line instead of a wreath type.

1   2   3

design recipe

4   By piping the modified connected reverse shells in alternating sizes of small and large, you can pipe five shells on a 15 cm round cake.

5   Start piping the upper row biased to the left. Pipe four small and large shells.

6   Pipe four shells also for the lower row the same way.

7   I will decorate it colorfully with purple pansies. When arranging flowers, start with the largest to set the anchor and finish with the smaller flowers.

8   Arrange the flowers naturally and finish with the silver leaf.

9   The cake with purple pansies is complete. You can also use stock flowers to decorate the same way (like the cakes on p.191 and 193).

tip

⑥ Piping should be started from the center-line so you can easily adjust the spacing between top and bottom, which will have a balanced look. The middle row will be piped skewed to the right side of the cake, and the other two to the left side.

⑦ You can decorate with yellow pansies and green thyme, or orange pansies and thyme also go well.

4   5   6

7   8   9

07

Basic cake finished with French star nozzle 2

Tool used
French star
867K

Technique | Piping hearts in single line		**Piping angle** | 45 degrees
Density | 85%		**Skill** | ●●●●●

ingredients & tool

mango, herbs, gold leaves, brush

design recipe

1   Prepare a 15 cm round cake, icing completed. `54p`

2   Start from 0.5 cm above the top of the cake, held at 90 degrees.

3   Hearts will be piped in a single line. Start piping from the top part of the cake. `124p`

tip

❸ For a 15 cm cake, use nozzle no.867K instead of 869K to cover the top completely. On the other hand, an 18 cm cake needs to be piped with 869K to cover the top.

1

2

3

design recipe

4 Start piping the second cream just above the tail of the first cream and go from left to right.

5 Continue to pipe in one row without making any gaps. For the last one, start from the left and pull the tail of the cream to the right.

6 Start piping the second line from right to left.

7 Pipe the second row without gaps as well. Like the first row, pipe the last cream from left to right.

8 Now the right side of the cake will be filled. Reverse all directions from the left side to make the left and right look symmetrical.

9 Pipe the third low without gaps as well, making it round and plump.

tip

4 The first starting point is essential for the single line heart piping. The decoration starts from the left side of the cake, and the piping directions proceed from left to right.

5 Conversely, piping from right to left may be unfamiliar, which can make the tail look awkward.

8 Start piping the right top side of the cake from left to right.

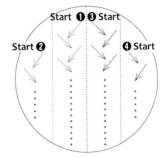

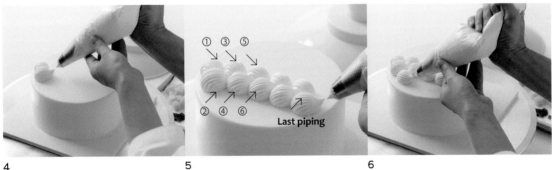

4   5   6

7   8   9

design recipe

10 Finish the fourth row the same way.

11 Place mango pieces cut into 1x1 cm cubes where the cream overlaps.

12 Place the mangos irregularly on top of the cake with the corners pointing up.

13 Put herbs where the mangos are so that the yellow and green colors can be seen at a glance.

14 Place the gold leaf on mangos to give accents.

15 The cake with gold leaves is complete.

tip

⑩ Overall, there should be no gaps, and the shape of the cream should be piped short and round to complete the cute decoration reminiscent of the knitted pattern.

10 11 12

13 14 15

08

Basic cake finished with French star nozzle 3

Tool used
French star
869K

Technique | Piping three hearts		**Piping angle** | 45 degrees	
Density | 85%		**Skill** | ●●●○○	

ingredients & tool

watermelon or mango, silver leaves, brush, fruit-baller

design recipe

1   Prepare a 15 cm round cake, icing completed. **54p**

2   I am going to use nozzle no. 869K to pipe three creams using the heart piping method. Start from 1 cm above the top of the cake, held at 45 degrees angle. **124p**

3   Start at the top left, pipe the first cream from left to right.

1   2   3

design recipe

4   Start the second cream right above the tail of the first cream and pipe from right to left. The third cream starts just above the second tail, pipe from left to right.

5   Pipe on the middle right part of the cake the same way.

6   Pipe on the bottom left part of the cake the same way.

7   I will decorate with the watermelon balls scooped out with a fruit-baller of 22 mm diameter.

8   Place the cute watermelon balls irregularly.

9   Put watermelon rind cut with the fruit-baller to add colors.

tip

❻ It's more effective to use a larger nozzle than a small nozzle when decorating partially.

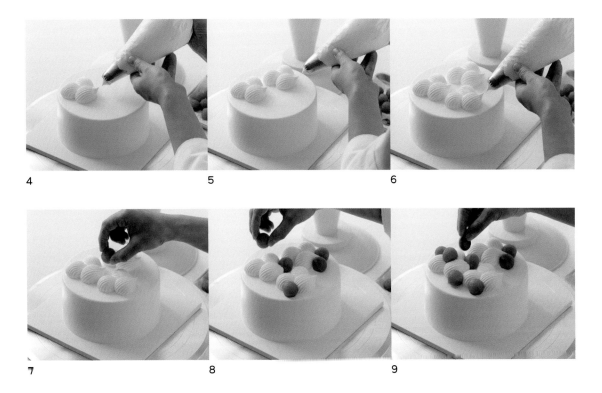

4   5   6

7   8   9

design recipe

10 Place the green rind next to the red watermelon to draw attention.

11 Stick silver leaf on the red watermelon to give a cool note.

12 The cake with watermelon is complete. You can also use mango and herbs to decorate the same way(as the cake on p.199).

10   11   12

09

Wreath style cake finished with ruffle nozzle

Tool used
Wilton 104

Technique	Connected small ruffled flowers	Piping angle	15 degrees
Density	85~90%	Skill	●●●●○

ingredients & tool

two different sizes of sugar pearls (arazans), gold leaves, brush

design recipe

1   Prepare a 15 cm round cake, icing completed. `54p`

2   I am going to use Wilton no.104 to pipe connected small ruffled flowers. Start from 0.5 cm above the top of the cake, held at 15 degrees angle, making sure not to open any wider. `130p`

3   Pipe three to four ruffles at once, then rotate the turntable counterclockwise.

1   2   3

design recipe

4   Pipe the cream while rotating the turntable to make natural little ruffles.

5   As you get closer to the finish, open the angle of the tip, taking care not to touch the previously piped cream with your left hand.

6   The small ruffles are finished. It looks more naturally bloomed when piped irregularly.

7   Place larger sugar pearls (arazans) on the ruffles.

8   Put the smaller sugar pearls (arazans) as well.

9   Finish by putting gold leaves on only partially to draw attention.

tip

❻ There are three types of Wilton 104 nozzles; Wilton, Wilton Korea, and Wilton China. Particularly, the Wilton China nozzle works well for expressing plump and cute ruffles.

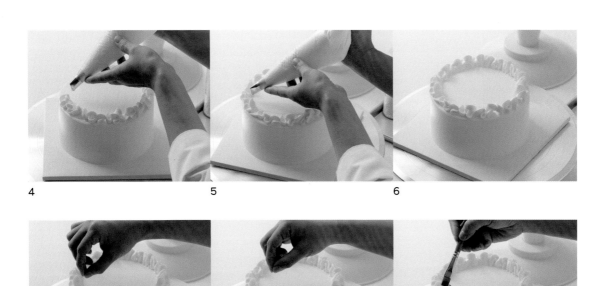

4   5   6

7   8   9

10

Square cake finished with ruffle nozzle

Tool used
Wilton 104

Technique	Connected small ruffled flowers	Piping angle	15 degrees
Density	85~90%	Skill	●●●●○

ingredients & tool

marigolds

design recipe

1   Prepare a 15 cm round cake, icing completed. `78p`

2   Start from 0.5 cm above the top of the cake, holding at 15 degrees angle; I will pipe ruffles continuously in one row from left to right. `130p`

3   Pipe a single row, continuously squeezing the cream, so the ruffles do not break.

tip

❸ Keep on squeezing while moving the hand in ∞ shape. If you move your hand up and down, it will pipe out like flat noodles. The same thing happens if you pipe from too high.

1   2   3

design recipe

4   In areas where it partially lacks volume, pipe in more cream to enhance the volume.

5   Repeat above; pipe a single line of connected ruffles, and fill in where it needs more volume.

6   The cake is completed filled with small ruffles. Ruffles look more naturally bloomed when piped irregularly.

7   Since the cake is gorgeous enough with just the ruffles, I will decorate only with small marigolds.

8   Arrange marigolds irregularly, between ruffles.

9   The cake is complete. The pattern cake filled with small ruffles is a fun cake that gives various looks depending on the direction you look at and when it's cut into pieces.

tip

❹ Before moving on to piping the next row, the part that lacks volume in the middle must be replenished to complete a cake with voluminous ruffles overall.

❺ If piped from too high, also the finished shape is not pretty. Make sure the distance between the cake and the nozzle doesn't go over 0.5 cm. Imagine that you are blooming small petals with your hand while piping.

❽ To prevent the volume of the ruffles from collapsing, pipe the cream as if to drop on top of the surface. To also show the sense of volume of the marigolds when viewed overall, the key is to lightly lay marigolds rather than pressing to stick it in.

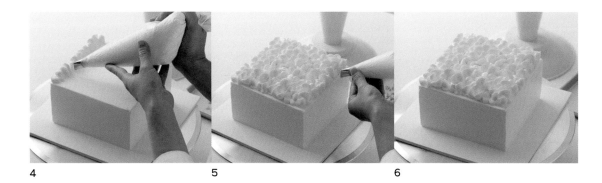

4   5   6

7   8   9

11

Square cake finished with V-shaped (St. Homoré) mozzle 1

Tool used
V-shaped 581
(St. Honoré)

Technique	Basic embossing	**Piping angle**	90 degrees
Density	80~85%	**Skill**	●●●●○

ingredients & tool

gold flakes, hazelnuts

design recipe

1   Prepare a 15 cm square cake, icing completed. `78p`

2   I am going to pipe voluminous basic embossing. Start from 1 cm above the top of the cake, held at 90 degrees angle. `141p`

3   Start piping from top left at 90 degrees.

tip

❷ It is common to start piping 1 cm above the top of the cake. If piped from too high up, you will pipe too much cream, and it will not look pretty. If you pipe at an angle of 45 degrees instead of 90 degrees from the beginning, the back of the cream will float from the surface not being able to fill in round shape.

1   2   3

design recipe

4　For a 15 cm cake, it is just enough to pipe six times per line.

5　Repeat the same way, piping one line at a time, not making any gaps.

6　Piping is complete. For a 15 cm cake, it must be completed with six creams horizontally and six vertically to fill the top.

7　Hazelnuts rolled in gold flakes will be used as decoration. Put gold flakes in a container with a lid.

8　Add roasted hazelnuts to the container, close the lid and roll it to coat the hazelnuts evenly.

9　It looks natural when the gold flakes are not coated too thick.

tip

❻ If there is an unusually large number of gaps after piping is done, or if the number of creams is 5x6, not 6x6, is because the angle of the nozzle is set narrower than 45 degrees.

Piped at 45 degrees: 6x6

Piped at below 45 degrees: 5x6

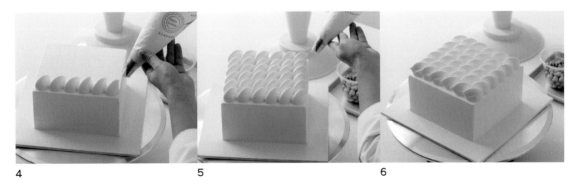

4　　　　　5　　　　　6

7　　　　　8　　　　　9

design recipe

10 Place the hazelnuts randomly on the creams.

11 The cake is completed. The decoration should feel like rolling waves.

tip

🔟 The hazelnuts rolled in gold flakes goes well with the winter theme. However, too much can be overwhelming, so use just enough to give a point.

10   11

12

Square cake finished with V-shaped (St. Homoré) mozzle 2

Tool used	**Technique** | Weaved embossing	**Piping angle** | 90 degrees
V-shaped 581 (St. Honoré)	**Density** | 80~85%	**Skill** | ●●●●○

ingredients & tool

mugwort powder, gold leaves, brush

design recipe

1   Prepare a 15 cm square cake, icing completed with whipped mugwort cream. `78p`

2   I will pipe voluminous weaved embossing. Start from 1 cm above the top of the cake, held at 90 degrees angle. `142p`

3   Start piping from top left at 90 degrees.

1   2   3

design recipe

4 Pipe 6 creams for a 15 cm cake.

5 To pipe in a weaved pattern, turn the turntable to change the direction after piping one row.

6 Try to pipe the second row without gaps.

7 Change the direction of the turntable back to where the first row was piped, and pipe the third row.

8 Six creams are piped using the same method. As we learned earlier on piping the basic embossing, the number of creams should be 6x6 to finish beautifully without any gaps. **212p**

9 The weaved embossing pattern is complete.

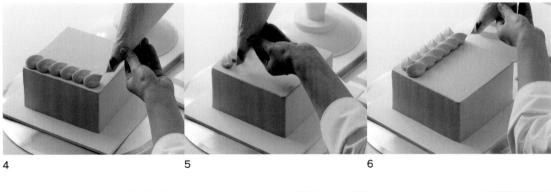

4   5   6

7   8   9

design recipe

10 The two corners facing each other will be decorated with mugwort powder. Prepare a fine sieve to sprinkle the powder.

11 Sprinkle the mugwort powder lightly.

12 I am going to place a gold leaf as a point, which goes well with mugwort.

13 Place the gold leaf on one corner of the cake.

tip

⓫ Be careful as it can look messy if you sprinkle too much.

⓭ For the embossing cakes, rather than deciding where the front will be before piping, it is better to complete the piping then find the prettier side by turning the cake around and check.

10　　　　　　　　11　　　　　　　　12

13

13

Chiffom cake finished with V-shaped (St. Homoré) nozzle

Tool used
V-shaped 682
(St. Honoré)

Technique	Connected large ruffled flowers	**Piping angle**	45 degrees
Density	85~90%	**Skill**	●●●●●

ingredients & tool

mugwort powder, gold leaves, brush

design recipe

1   Prepare a 15 cm square cake, icing completed with whipped mugwort cream. **72p**

2   Point the open end of the V-shape face outside, hold the nozzle 1 cm above the top of the cake, angled to 90 degrees.

3   Pipe the cream in a round shape while rotating the turntable.

tip

❷ If the nozzle's open end is facing up, it's difficult to create natural petal shapes. Do remember that the open end of the nozzle should face outside.

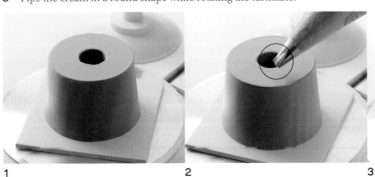

1   2   3

완두콩맘

design recipe

4   By adjusting the piping speed, rotate the turntable with the left hand and pipe with the right to make a large ruffle-shaped flower.

5   Make sure the outermost part of the piped cream fits the edge of the cake as much as possible.

6   Piping is completed. When removing the nozzle after piping is done, relax your hand and pull the nozzle toward the hole so that the horn of the cream does not rise.

7   The ruffle itself is gorgeous enough as a design so it will be decorated only with gold leaf.

8   Arrange gold leaves randomly.

9   Finish by placing a cake topper at the point where the beginning and the end of the piped cream meet.

tip

❺ If the cream is piped too inside from the outer edge of the cake, the ruffle becomes smaller because of the hole. Try to pipe along the outer edge as much as possible.

4   5   6

7   8   9

14

Chiffom cake fimished with two mozzles

Tool used
Round 809,
French star 867K

Technique	Teardrops	Piping angle	90 degrees
Density	80~85%	Skill	●●○○○

ingredients & tool

isomalt chips, silver leaves, brush

design recipe

1   Prepare a 15 cm chiffon cake, icing completed with whipped black sesame cream. **72p**

2   Fill a piping bag fitted with the round nozzle and the whipped black sesame cream. Hold the nozzle 1 cm above the top of the cake, angled to 90 degrees to pipe teardrops. **117p**

3   Be careful not to move your hands, and when the cream touches the cake, pull it up quickly.

1   2   3

design recipe

4   Pipe 1~2 mm away from the edge of the center hole. Also, stay 1~2 mm in from the outer edge of the cake and pipe the teardrops irregularly.

5   In a piping bag fitted with the French star nozzle, fill it with plain whipped cream. Hold the nozzle 1 cm above the top of the cake, angled to 90 degrees to pipe. **121p**

6   Pipe the plain cream around the black sesame cream to go along with the previously piped cream.

7   The piping is finished with some empty spaces so that it doesn't look crowded.

8   Among the prepared isomalt chips, choose the largest and most splendid looking chip and insert as a pivot.

9   Insert a smaller isomalt ship in front of the previously inserted chip.

tip

❹ If the cream is piped too close to the center hole, the cream may get in the hole and may look stuffy overall. Likewise, if it is piped too close to the outer edge of the cake, the cream may fall onto the side of the cake.

❺ The creams with particles can get in the way of piping, such as black sesame cream or Oreo cream, are easier to pipe with plain round nozzles rather than a French star nozzle with multiple points.

❻ When using both round and French star nozzles, it is better to pipe with the round nozzle first. This makes it easier to pipe with the French star nozzle, which is relatively more comfortable to handle, into the empty areas.

4   5   6

7   8   9

10   Place another small isomalt chip on the opposite side, so it doesn't get biased to one side, and finish with silver leaves.

11   The cake is finished with a little bit of silver leaf.

10   11

15

Oreo dome cake finished with straight edged dome cake scraper

Tool used
Plain edge
dome scraper

Technique | Dome icing with building a crown   **Skill** | ●●●○○

ingredients & tool

mini size Oreo cookies, silver leaves, brush.

design recipe

1   Put Oreo cookies in a piping bag and press with a rolling pin, crush them fine to make Oreo icing cream. `35p`

2   Use the plain edge dome scraper to ice a 15 cm round cake, and build the crown naturally. `62p`

tip

❶ If the cookie particles are too thick, it may get in the way of the scraper while icing and create lines.

❷ The reason for building the crown is to fill the ganache stably while making the cake look taller.

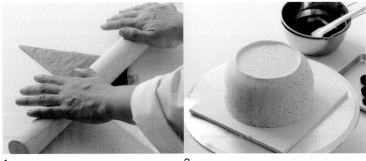

1   2

완두콩맘

design recipe

3   Pour 30 g of ganache (fresh cream and dark chocolate mixed 1:1 ratio) that was on a double boiler in the center of the cake.

4   Pour until the ganache reaches the edge of the crown evenly.

5   Store the finished cake in the refrigerator for a while to let the ganache harden a little.

6   First, fix the largest Oreo ornament as a pivot.

7   Place the rest of the other ornaments accordingly.

8   Finish with silver leaves on Oreo and the ganache.

tip

❹ If you pour the ganache as you move your hand back and forth, the iced whipped cream may melt and float on the surface of the ganache. Therefore, pour from not too high, without moving.

❻ Making Oreo ornaments
① Prepare mini Oreo cookies, melted chocolate, and short and thin toothpicks.
② Dip the toothpick in the melted chocolate, stick it in the Oreo, and coat the connecting part with the melted chocolate as well.
③ Connect the Oreos in Mickey Mouse or cactus shape, then freeze on a flat tray. The frozen chocolate in between will work as glue.
④ Use the frozen ornaments as decorations on the cake.

3   4   5

6   7   8

16

Dome cake finished with ruffle nozzle

Technique	Small ruffled flowers	**Piping angle**	15 degrees
Density	85~90%	**Skill**	●●●●○

Tool used
Wilton 104

ingredients & tool

edible flowers, silver leaves, brush

design recipe

1. Prepare a 15 cm dome-shaped cake, icing completed with whipped strawberry cream. `62p`
2. I will pipe small ruffled flowers. Hold the nozzle 0.5 cm above the top of the cake, angle not to exceed 15 degrees to start. `129p`
3. First, bloom three petals, and then two petals. Move up and down to pipe the cream to make small petals bloom.

1   2   3

완두콩맘

design recipe

4   Pipe small petals without any gaps. Fill in the spaces while piping to make it more voluminous.

5   The piping is finished, filling the gaps along the way.

6   Decorate with edible flowers that go well with the cake.

7   Place silver leaves randomly.

8   Stick a clean looking rectangular cake topper on the decorated part to draw attention.

9   The cake is complete.

tip

❺ If you wrap the top completely with a wreath shape, it can look stuffy. I recommend starting at 5 o'clock location and end at 12 o'clock. That way, when viewed from the front after completion, the front will be ruffles and the back will be decorated with petals- the tail of the cream won't be visible, which will look more natural.

4   5   6

7   8   9

17

Mango ganache drip cakes

Technique | Center drip	**Skill** | ●●●○○

ingredients & tool

mango, edible flowers, herbs, silver leaves, brush, fruit-baller

design recipe

1 Finish the icing on the tall 15 cm round cake, and prepare mango
ganache. `54p`

2 Prepare to pour the ganache, 1 cm inward from the edge of the cake.

3 Pour the ganache from the starting position while slowly moving towards
your body.

tip

❶ Mix mango puree and white
chocolate in a 1:1 ratio to make the
mango ganache.

1   2   3

design recipe

4   Continue to slowly pour the ganache towards the front, stop at 1 cm before the edge of the cake, and let it flow down by itself.

5   The ganache has dripped down the side of the cake. The ganache looks natural and stylish when it touches the bottom with one line on the back of the cake and one on the front.

6   After the ganache sets slightly, top with mangos scooped out with a fruit-baller.

7   Do not place mangos the way it will distract attention. 6~7 pieces are suitable for a 15 cm cake.

8   Decorate with pansies where the mangos are. Set a guide with the largest flower.

9   Arrange smaller flowers as well.

tip

❹ Drizzle a line with a spatula to check the viscosity of the ganache. It is ready to use when the lines appear and disappear soon. If the line does not fade away and the ganache gets stacked, it is too thick.

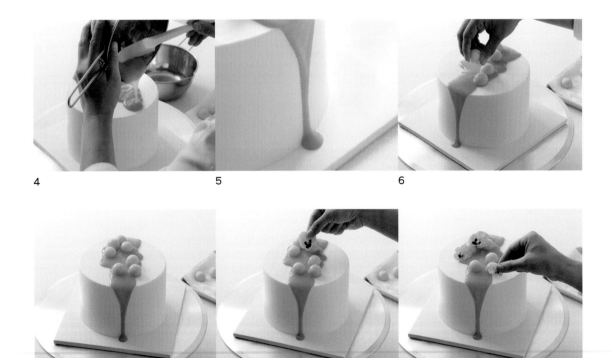

4   5   6

7   8   9

design recipe

10 Add extra color with herbs where the flowers are placed.

11 Accentuate with gold leaf that goes well with the mango.

12 The cake is complete.

10   11   12

18

Two types of tree cakes

Tool used
Plastic card

Technique | Cake combing, double icing   **Skill** | ●●●●●

ingredients & tool

Two different sizes of sugar pearls (arazans), decosnow, chocolate decorations

design recipe

1   Stack 7 sizes of round genoise sheets in diameters of 15, 13, 11, 9, 7, 5, and 3 cm in order.
     Ice all together into a cone shape.

2   Lines will be drawn on the side of the cake using a 3x5 cm plastic card. Rotate the turntable once or twice as is so the
     bottom can be organized.

1   2

design recipe

3 Do not open the angle of the card between the cake board by more than 15 degrees. With the edge of the card touching the surface of the cake, rotate the turntable and make a line by moving your right hand up from the bottom to top.

4 Make the line while giving a steady pressure with your hand.

5 Check to see that the sideline is made thin and round.

6 Hold the card straight as it rises to the top, and finish with only the tip of the card is touching.

7 The sideline is complete.

8 Decorate with larger size sugar pearls (arazans).

tip

❹ If the spacing of the line is even, it gives a neat look, and if it's irregular, it feels rhythmic, finishing it with a fun look. But, if the spacing is too narrow, it can look stuffy.

❺ If too much pressure is given, it will peel off the cream, and if the card is bent too much, it can create layers, and the cake sheets will show.

❼ If your hand is shaking a lot, the line will be uneven. If so, attach your arm to your torso so that your arm does not shake while working.

❽ Stick the sugar pearls (arazans) by pressing it lightly just enough so that it does not fall off the cream. Be careful as the cream may get on your hand and damage the icing. It may be easier to put cream on a long skewer, such as toothpicks, and then pick up the sugar pearls (arazans) to stick on the cake.

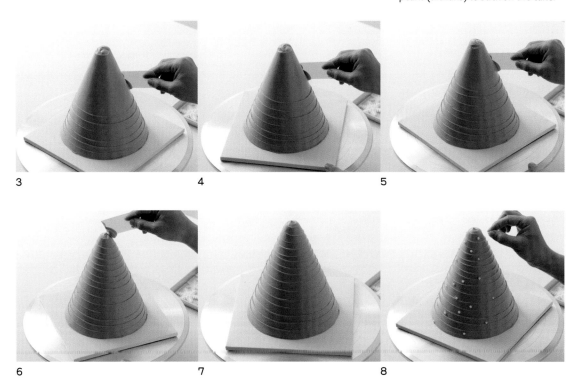

3   4   5

6   7   8

design recipe

9　Place snowflake-shaped chocolate decorations in between.

10　Top with the largest snowflake decoration on the very top.

11　The first tree cake is complete.

12　The second design is finished with the double icing method. 286p

tip

⑫ After completing the double icing, decorate it with different sizes of arazans, and sprinkle a little bit of decosnow to give it a snowed look.

9

10

11

12

19

Heart cake finished with V-shaped (St. Homoré) nozzle I

Tool used
V-shaped 25
(St. Honoré)

Technique	Large ruffles	**Piping angle**	45 degrees
Density	85~90%	**Skill**	●●●●○

ingredients & tool

gold leaves, brush

design recipe

1   Prepare a 15 cm heart-shaped cake, icing completed. **82p**

2   I will pipe large ruffles. Start from 1 cm above the top of the cake, turn the opening to face up, and hold at 45 degrees angle.

3   As if drawing a small 8, pipe small at first only by flicking the wrist.

tip

❷ Turn the cake around to pipe- this way, the ribbon looks naturally laid back when viewed from the front.

1   2   3

design recipe

4   Pipe while increasing the length and width of an 8.

5   From the middle point, gradually reduce the size of an 8 to finish.

6   Piping is finished.

7   Randomly arrange gold leaves to decorate.

8   The cake is complete.

tip

❻ When decorating the left side of the cake after icing the heart-shaped cake, make sure that the finishing line on the top of the cake is gathered to the left as much as possible. That way, the finishing line can be covered with the piped decoration, making a high-quality cake.

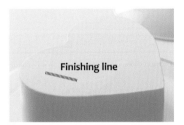

Finishing line

4   5   6

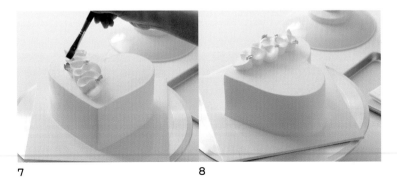

7   8

20

Heart cake finished with V-shaped (St. Homoré) nozzle 2

Tool used
V-shaped 25
(St. Honoré)

Technique | Basic embossing	**Piping angle** | 90 degrees
Density | 80~85%	**Skill** | ●●●●○

ingredients & tool

edible roses, arazans

design recipe

1   Prepare a 15 cm heart-shaped cake, icing completed. `82p`

2   I will pipe basic embossing irregularly. Start from 1 cm above the top of the cake, holding at 45 degrees angle.

3   Starting from the top left of the cake, pipe the cream while moving downwards.

1   2   3

design recipe

4   Take care not to make big gaps between creams while piping.

5   It looks more natural to pipe slightly irregularly then making it uniform.

6   Piping is complete.

7   Edible roses will be used to decorate to give more volume.

8   Arrange the roses irregularly as well for more natural look.

9   Finish with sugar pearls (arazans) to give a lovely mood.

tip

❻ When decorating the left side of the cake after icing the heart-shaped cake, make sure that the finishing line on the top of the cake is gathered to the left as much as possible. That way, the finishing line can be covered with the piped decoration, making a high-quality cake.

Finishing line

4   5   6

7   8   9

21

Lime pattered cake finished with V-shaped (St. Homoré) nozzle

Tool used
V-shaped 25
(St. Honoré)

Technique | Line pattern	**Piping angle** | 45 degrees
Density | 80~85%	**Skill** | ●●●●●

ingredients & tool

gold leaves, brush

design recipe

1   Prepare a 15 cm round cake, icing completed, with a crown built on top. `54p`

2   Long continuous lines will be piped. Start low, 0.5 cm above the top of the cake, holding at 45 degrees angle.

3   It is crucial to pipe the line boldly at once so the lines do not wobble.

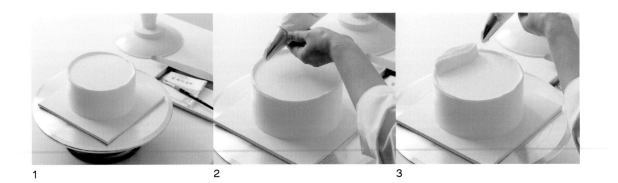

1   2   3

design recipe

4 It looks beautiful when the lines are curved in the shape of waves.

5 Start piping the line from inside the crown, and go slightly outside the crown to finish.

6 Organize the side of the cake and the crown with a spatula.

7 Place gold leaves as point decoration and finish.

tip

❹ It is a design that the line of the creams can be seen differently depending on the direction of the shadow and the viewing angle.

❻ Finish with a spatula, ensuring that the ends of the piped lines are neatly organized.

❼ The most important point for completing a smooth line is 'density.' The density should be a little thinner than the usual piping cream so that the line does not break while it's being piped to make a natural curve.

4   5   6

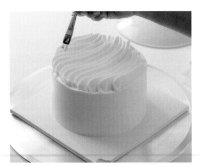

7

22

Chiffom cake finished with scooping technique

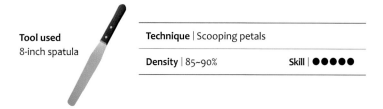

Tool used
8-inch spatula

Technique | Scooping petals	
Density | 85~90%	**Skill** | ●●●●●

ingredients & tool

gold leaves, brush

design recipe

1   Prepare a 15 cm chiffon cake, icing completed. `72p`

2   I will scoop neat petals. Even out the texture of the cream, stick the end of the spatula into the cream diagonally and pull out. `152p`

3   The cream should look like a pretty petal when the spatula is pulled out.

tip

❸ If the density of the cream is too thin, the tip of the petal may come out sharp but will fall back, too thick, then the tip may come out rounded.

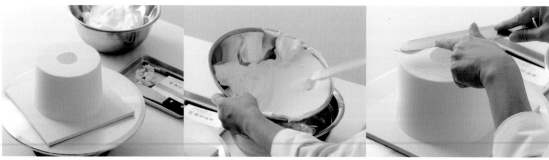

1   2   3

design recipe

4   Do not open the angle of the spatula excessively, while taking carew not to let the tip of the petal come with the spatula and pull out towards the hole.

5   Apply the second petals slightly overlapping the first.

6   Place the third petal the same way, while checking if the amount of the cream is right.

7   Press gently with a spatula where it slightly overlaps the second petal.

8   When pulling out the spatula, be careful not to touch the top of the opposite side of the hole.

9   The petals are complete.

tip

❺ The petals look much more natural when they overlap slightly.

❻ Be careful not to go too far off the outer edge of the cake when pressing the cream out, to prevent the petals from bending over the outside of the cake.

❾ For a 15 cm chiffon cake, eight to nine petals are suitable. Too many petals may look stuffy.

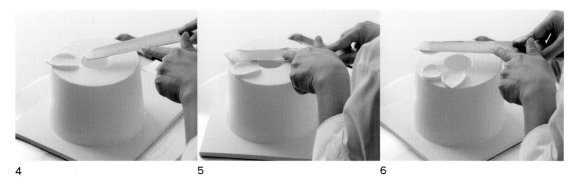

4   5   6

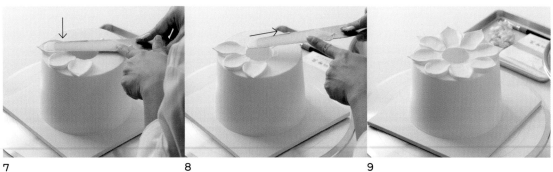

7   8   9

design recipe

10 When the petals are done, arrange gold leaves or edible flowers to decorate.

11 The cake is complete.

10   11

23

Pumpkin dome cake finished with a plastic card

Tool used		Technique	Dome icing- wave pattern	Skill	●●●●●
Plastic card					

ingredients & tool

marigolds

design recipe

1 Prepare a 15 cm dome-shaped cake, icing completed with whipped sweet pumpkin cream using a plastic card. 62p

2 I am going to make pumpkin ribs. Bend the card according to the dome shape and wrap it completely.

3 Make sure that all the curved parts of the card touch the cream so that you can pull the cream out without any gap.

1   2   3

design recipe

4 With a little force on the hand, ice the cake by moving the card like a wave to create the pattern.

5 Trim the shape until it looks like the ribs of a pumpkin, and tidy up the cream on the bottom of the cake.

6 The icing is complete.

7 I will decorate with marigolds that goes well with the yellow sweet pumpkin cake and make the curved design stand out.

8 Arrange the marigolds irregularly on the top and side of the cake.

9 The cake is complete. It goes well with gold leaves also other than marigolds.

tip

❹ Be careful as the dome shape may get damaged if the fingers holding the card moves.

❾ To make the ribs stand out more, apply a little more strength when icing with the card. By tilting the angle of the card, you can complete a different line pattern. Just by changing the height of the cake, it will have a different impression.

4   5   6

7   8   9

24

Melting cake using fresh cream

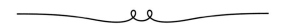

Technique	Dripping	Skill	●●●●○

ingredients & tool

Pasteur brand 45% fat dairy fresh cream

design recipe

1 Prepare a 15 cm dome-shaped cake, icing completed with whipped raspberry cream. `54p`

2 Mix 70 g of 45% dairy fresh cream with 10 g of sugar and 18 g of mascarpone into a thick consistency to prepare the melting cream.

3 Without moving the position, pour the cream only in the center of the cream.

tip

❷ To make the melting cream, use cream with a milk fat content of 45%, which has high viscosity for stability, and it works well to keep the dripped shape after completion.

1   2   3

design recipe

4   Pour the cream until it fills 1~2 cm from the outer edge of the cake.

5   I will work on the top without opening the spatula's angle too wide.

6   Rotate the turntable while relaxing the hand as much as possible on the hand holding the spatula. If too much pressure is given on the hand while rotating, you might touch the already iced cream.

7   With your hand relaxed, pull the spatula out.

8   Gently tap the cake board with your hand so that the cream can flow down naturally.

9   Place a cake topper to finish. A simple rectangular topper goes well.

tip

❹ Pouring too much cream can cause a lot of cream to drip down and look messy.

❻ If you rotate too much, the iced cream on the cake may be pressed flat, or the color may get mixed. It's best to finish within two turns.

4   5   6

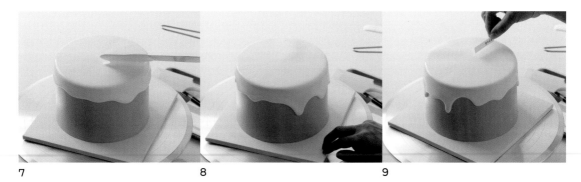

7   8   9

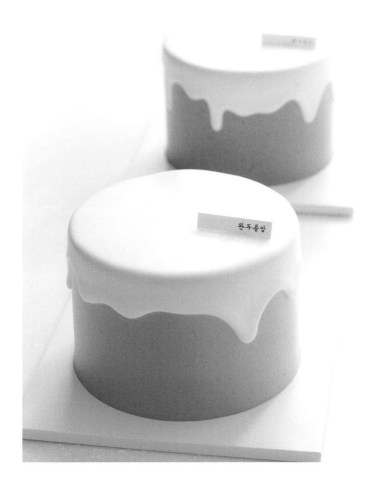

25

Chiffom cake decorated With various fruits

Skill | ●●●○○

ingredients & tool

mini apples, strawberries, blueberries, raspberries, plastic chocolate, silver leaves, brush

design recipe

1 Prepare a 15 cm chiffon cake, icing completed with whipped strawberry cream. **72p**

2 I am going to arrange the fruits by turning the cake in one direction.

3 Place the larger fruit first to set as an anchor and then put smaller fruits in several places. Mini apples look much better with the stems on because it creates high and low when the cake is completed.

1   2   3

design recipe

4   Place blueberries randomly as well. Place blueberries halved with flesh and seed visible, also the whole fruit; it looks prettier.

5   When arranging the fruits, it is better to decorate in one direction completing bit by bit rather than placing them here and there.

6   Place raspberries in between.

7   All the fruits are arranged. Try to fill with no empty areas.

8   Place flower decorations made of plastic chocolate to accentuate.

9   Use different size flower decorations to decorate it cute.

4   5   6

7   8   9

design recipe

10 Put silver leaves on the red fruits.

11 The cake is complete. The extended stems of the mini apples give a sense of movement.

tip

⑪ If you use the strawberries without its green leaves, use green plants such as mint or thyme to give them a color. When the red fruits are used as the main ingredients, it looks pretty to provide the green color.

10   11

26

Chiffom cake finished with round nozzle

Tool used
Round 809

Technique	Laid piping	Piping angle	45 degrees
Density	80~85%	Skill	●●●●○

ingredients & tool

edible flowers, herbs

design recipe

1 Prepare a 15 cm chiffon cake, icing completed. `72p`

2 Hold the nozzle 0.5 cm above the top of the cake, angled to 45 degrees to pipe teardrops. `118p`

3 Without moving your hands, pipe as if you are pushing in only the cream and then pull out the nozzle.

tip

❸ When pulling out, make sure the tail of the cream is pointing to the center of the cake so that it becomes a lying teardrop shape.

1

2

3

design recipe

4 Pipe 1~2 mm inside from the outer edge of the cake.

5 For a 15 cm chiffon cake, piping ten creams should fill without gaps.

6 Piping is complete.

7 I am going to decorate with small edible flowers that match the cute teardrop creams.

8 Gather the flowers on the top left of the cake to draw attention, and put one more on the bottom right as a point.

9 Make the flowers stand out by placing herbs around them.

tip

❹ If you pipe from too high or the angle is narrower than 45 degrees, the cream will come out longer. The piped cream looks pretty when it is finished in cute teardrops, with a short and plump shape.

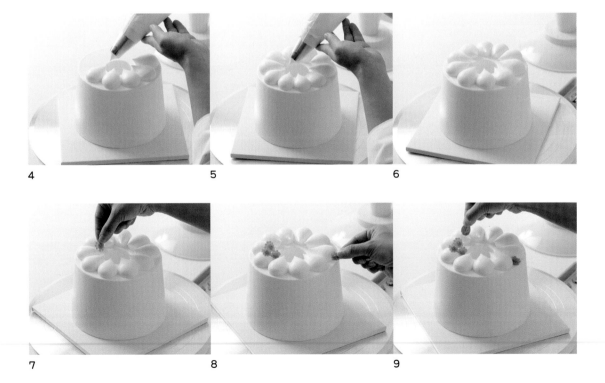

4   5   6

7   8   9

design recipe

10 Add herbs to the flower decoration at the bottom left of the cake.

11 The cake is complete.

10   11

27

Chiffon cake finished with two-toned icing

| **Technique** | Two-toned icing | **Skill** | ●●●○○ |

ingredients & tool

mini pinecones, pomegranate, decosnow, dill herb, silver leaves, brush

design recipe

1   Prepare a 15 cm chiffon cake. Ice the lower part with the Oreo whipped cream and the upper part with white whipped cream to make it look like snow is piled up. **72p**

2   When the icing is complete, sprinkle decosnow to make it feel like snow has fallen.

3   Mini pinecones will be used to decorate.

1   2   3

design recipe

4   I am going to decorate only on one side. It also looks good to put it all around.

5   Place the red pomegranate seeds in between the pinecones to accentuate.

6   Use dill to decorate, one of the herbs that go well with pinecones and winter theme. Rosemary also goes well.

7   Put silver leaves on the red fruits to finish.

tip

❺ You can only use pomegranate seed to decorate, or use it in a lump after splitting it carefully.

❼ For two-toned icing and double icing, it looks more balanced to ice the two creams exactly in half. If the ratio of the cream on the top or bottom is larger, the overall balance will be off and may look unstable.

4   5   6

7

28

Tall cake finished with rectangle cake comb

Tool used	Technique	Cake comb icing	Skill	●●○○○
Cake comb				

ingredients & tool

ganache, gold flakes, brush

design recipe

1   Ice a tall round cake with Oreo whipped cream, and use a cake comb to finish the sides. `54p` `164p`

2   The top is going to be filled with ganache, so build the crown on top.

tip

❶ The cake combs have deep grooves, so you need to apply a generous amount of cream to ice, to prevent the cake sheets from showing.

❷ After the icing is finished, keep it in the refrigerator for 10 to 20 minutes and prepare 1:1 ganache (30 g fresh cream, 30 g dark chocolate).

1   2

design recipe

3   Without moving, pour the ganache only in the center.

4   The ganache is filled.

5   For cakes with a lavish side, it's better to decorate the top simple. This time, I am going to finish it only with gold flakes.

6   Use a brush to dust the gold flakes evenly.

7   The cake is complete.

tip

❸ If you move while pouring the ganache, the cream on the top of the cake may float on the surface of the ganache, creating spots.

❻ Flicking on your fingertip will help to sprinkle evenly.

3   4   5

6   7

29

Pattern cake finished with spatula

Tool used
Spatula

Technique | Stripes   **Skill** | ●●●●○

ingredients & tool

mini apples, green leaves

design recipe

1   Prepare a 15 cm round cake, icing completed with the raspberry whipped cream. **54p**

2   I am going to lower the angle of the spatula to about 30 degrees and create lines with the hands relaxed.

3   Make lines in uniformed intervals.

tip

❷ Be careful; if you open the angle more than 30 degrees, the iced cream on the top of the cake will shave off while making the lines.

1   2   3

design recipe

4   Make lines by controlling the strength of your hand, firm and soft, while the spatula blade stays on the cream.

5   Be careful not to make the edge collapse as the spatula gets closer to the outer edge as it moves down. When making the last line, try to finish inside the outer edge of the cake.

6   When the pattern is complete, use the spatula to ice the sides once more to build the crown.

7   A low and natural crown is built.

8   I will decorate with small and cute mini apples.

9   Use different sizes of mini apples to create a rhythmic look and make the finished look more beautiful.

tip

❹ If the spatula touches and lifts from the cream when marking the lines, an empty space is created, making another line.

❺ Finishing the line too close to the outer edge can cause the cream to sag on the outside.

❻ It is a process of reinforcing once more by building the crown to fill the empty part of the line.

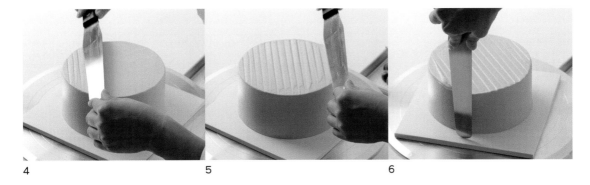

4   5   6

7   8   9

10 Arrange green leaves around the mini apples to give a point.

11 The cake is completed cute by accentuating with the mini apples' stems intact and green leaves.

⑩ The key is to use the backside of the leaves that are lighter in color than the darker front side; the veins are more visible so that the curved texture can be expressed more.

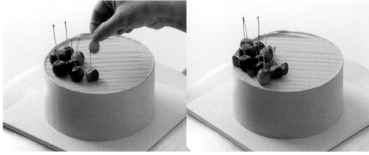

10   11

30

Round cake finished with double icing

Tool used
Spatula

Technique | Double icing   **Skill** | ●●●●●

ingredients & tool

rose petals, mirror glaze (decogel), gold leaves, brush

design recipe

1   Prepare a 15 cm round cake, icing completed with the white (plain) whipped cream. **54p**

2   I am going to double-ice with raspberry whipped cream. Place the amount of cream to cover half the height of the cake.

3   Spread the cream with your hand relaxed.

1   2   3

완두콩맘

design recipe

4   The cream should spread out to the sides to be able to ice half the height of the cake.

5   Align the tip of the spatula blade in the center and slightly open the angle. Keep your hand relaxed the whole time. Be careful not to rotate the turntable too much; otherwise, the cake sheets will show.

6   I will build a crown on top.

7   Place only the tip of the blade in contact with the raspberry whipped cream, then rotate the turntable and slightly spread the cream to build the crown.

8   Use the spatula to remove the cream a little from the side where it has a lot of raspberry cream, leaving only the amount that will cover half the height of the side.

9   Start icing the side with your hand relaxed as much as possible.

tip

❹ When double-icing, it is better to apply the cream on top of the cake as thin as possible. If the cream on the top part gets thicker, the moment it is pushed in with a spatula, the white cream gets pushed under the raspberry cream and the two colors will get mixed.

❾ If you hold the spatula strong when icing, the white cream and raspberry cream will blend and become two-toned icing instead of double icing.

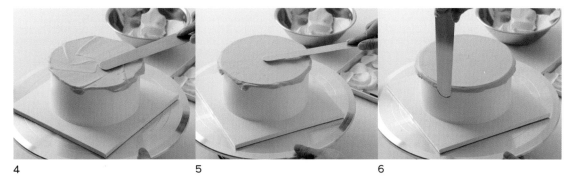

4   5   6

7   8   9

design recipe

10 The point is to hold the spatula straight down, not to touch the white whipped cream but only touch the raspberry whipped cream.

11 Ice to cover half of the side.

12 It is okay to finish with the crown built naturally. Here, I will trim the crown to finish neatly.

13 Open the tip of the spatula blade and finish the top neatly.

14 I am going to decorate it with a light pink rose to create a lovely cake.

15 Place the petals on the top left side of the cake for a gorgeous yet elegant look.

tip

⑩ Finish icing the side of the raspberry cream at the same position as the finish line of the white cream on the side. It's easier to decide where the front of the cake will be, only when the finish lines match.

⑪ For the double icing, it looks most stable to have equal parts of two creams on the side of the cake. If one of the creams is used more, the overall balance and proportions can look unstable.

10   11   12

13   14   15

design recipe

16 Attach a rose petal on the bottom of the cake as a point.

17 Lightly pipe the mirror glaze (deco gel) on the petals to create a dewy effect.

18 Stick gold leaves over the mirror glaze (deco gel).

19 Place a neat rectangular cake topper to finish.

tip

⑰ Put the glaze in a small piping bag, cut the tip with scissors, and pipe in small portions to use.

16　　　　　17　　　　　18

19

31

Children's cake finished with sprinkles

Tool used Round (14 mm diameter)	**Technique** | Sprinkle icing	**Piping angle** | 90 degrees
	Density | 80~85%	**Skill** | ●●○○○

ingredients & tool

sprinkles, chocolate decotations

design recipe

1 Prepare a 15 cm round cake, icing completed with whipped cream mixed with ground sprinkles. **54p**

2 Put the largest chocolate decoration that will be the centerpiece.

3 Arrange the rest of the chocolate decorations.

tip

❶ If the sprinkles are mixed with the whipped cream without grinding, the thick particles will scratch while icing. After whipping the fresh cream, lightly mix the finely ground sprinkles and start icing.

1   2   3

design recipe

4   I am going to decorate in the center of the cake, but it looks prettier to arrange irregularly than uniformed.

5   Use a small round nozzle to pipe small and cute creams randomly to look like a forest. **117p**

6   Try not to pipe too much so that the animal friends can be the main characters.

7   Piping is complete.

8   Sprinkle some sprinkles on the piped cream to give more colors.

9   The cake is complete. It is a simple and easy cute cake.

tip

❽ The Nonpareils (round shape) sprinkles look better than the Jimmies (rod shape).

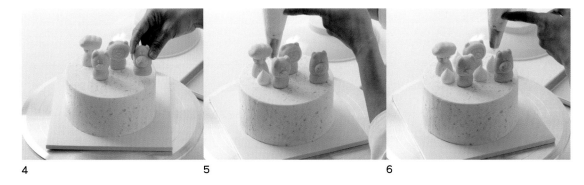

4   5   6

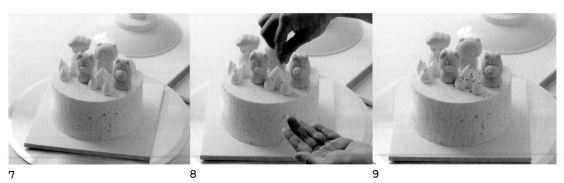

7   8   9

32

Round cake finished with two types of nozzles

Tool used
Round 809,
French star 869K

Technique | Teardrops		**Piping angle** | 90 degrees	
Density | 80~85%		**Skill** | ●●●○○	

ingredients & tool

banana slices, banana croquant, flower shape cutter, gold leaves, brush

design recipe

1   Prepare a 15 cm round cake, icing completed with ganache whipped cream, and prepare decoration creams (ganache whipped cream, white whipped cream). `54p`

2   Prepare to pipe, holding the nozzle 1 cm above the top of the cake, angled to 90 degrees. `117p`

3   Using the round nozzle, pipe two equal sized teardrops of the ganache whipped cream, 1~2 mm from the edge of the cake.

1   2   3

design recipe

4   Piping with the round nozzle is finished.

5   In the empty spaces, pipe the white whipped cream in a small and cute shape with the French star nozzle. **121p**

6   Piping is finished using two different nozzles.

7   Cut the banana slices with the flower-shaped cutter to make banana flowers.

8   Over ripen banana will get squished and won't cut out pretty. Use underripe bananas for better shape.

9   I am going to decorate the left side of the cake.

tip

❹ When piping with both round and French star nozzles, it is manageable to pipe with the round nozzle first, which is relatively difficult to pipe, then finish with the French star nozzle.

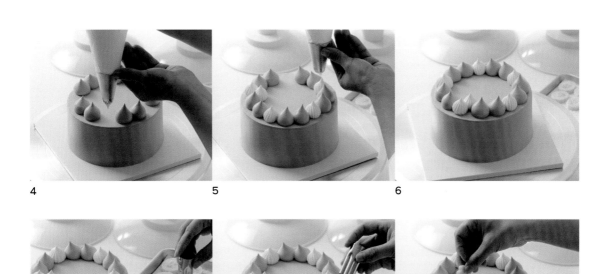

4

5

6

7

8

9

design recipe

10 Do decorate around the left side of the cake, but put one piece of the banana on the right side to make a point decoration.

11 I will add banana croquants. Choose a large and pretty piece and center it.

12 Put one piece of the banana in front of the croquant, and place a smaller croquant in the front.

13 Top with a gold leaf on the croquant to finish.

10 11 12

13

33

Square cake finished with big ribbon pattern

Tool used
Petal
125K

Technique | Big ribbons	**Piping angle** | 15 degrees
Density | 85~90%	**Skill** | ●●●●○

ingredients & tool

cornflowers, silver leaves, brush

design recipe

1. Prepare a 15 cm square cake, icing completed with raspberry whipped cream. **78p**

2. Since the opening of the nozzle is quite large, prepare the cream much thicker than usual. Test to see whether the ridges of the ruffles show clearly.

3. Prepare to pipe, holding the nozzle lower than 0.5 cm from the top of the cake, angled to 15 degrees. **136p**

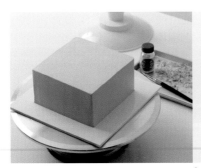

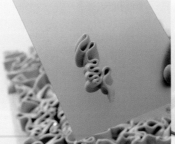

1

2

3

design recipe

4   When piping the large ruffles, you should pipe it low, very close to the surface; otherwise, the cream may flip when piped from too high.

5   Pipe out the cream vertically without stopping.

6   It is essential not to move your hand side to side while piping but to pipe it so that the ruffles are naturally formed while the cream is being piped out.

7   For a 15 cm cake, it looks best to finish piping in 5~6 rows.

8   Arrange the subtle pink-toned cornflowers that match the pink ruffles.

9   Place silver leaves randomly to finish.

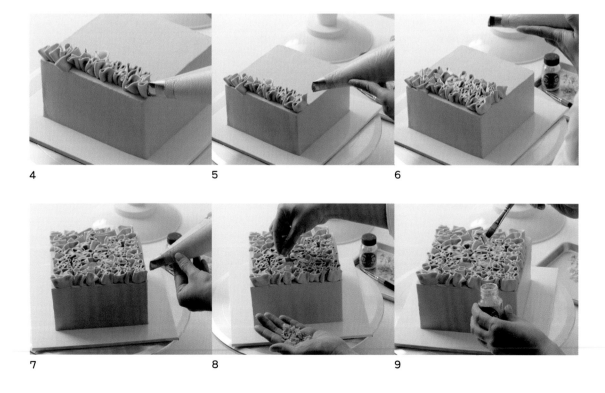

4   5   6

7   8   9

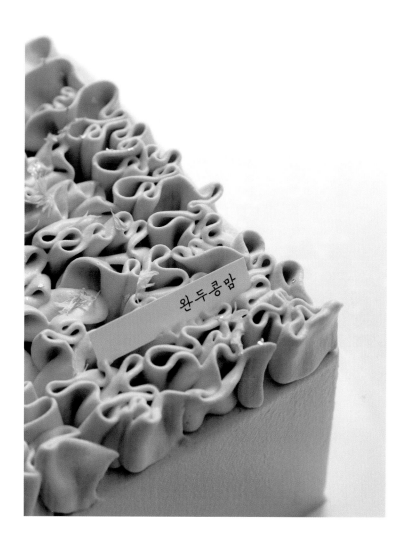

완두콩맘

Lined Dome Cake

It is a dome cake made by making the lines by moving a small plastic card upward, finishing with a naturally raised crown. Arrange rock chocolates and microwaved sponge decorations to create a unique texture, then finish with large gold leaves to make a luxurious design. When pulling up the plastic card, it is important to use minimal strength so that the side of the cake is not carved excessively.

완두콩맘

Tambourine Cake

This is a cake I made in response to an acquaintance's accidental question about why there is a triangle cake but no castanets or tambourine cakes. After icing with a ripple edge dome scraper, I made pockets on the side of the cake with the residual heat of a hot measuring spoon and filled it with a ganache. Depending on the color of the ganache, and the size of the pockets made by the measuring spoons, you can create various designs. Icing with the ripple edge dome scraper is the best way to enhance the convex texture the most after making the pockets.

Pumpkin Matcha Cake

It is a cake designed in the shape of a sweet pumpkin that reminds me of autumn. The dome icing is done with the sweet pumpkin whipped cream, and the side of the cake is patterned like the ribs of the sweet pumpkin and accessorized with honeycomb decorations and a gold leaf. Dome icing can be applied in various ways, such as the roof of a palace, Kisses shape, apple, etc. The conventional tools are great, but if you can get used to tools, such as plastic cards, you can make desired shapes, making cake designing even more fun.

Barefoot Dome Cake

When I was teaching at a chocolate studio, I wanted to design a simple but more interesting dome cake and made this design. Thinking that the dark brown color of the ganache showing on the bottom of the cake would look pretty, I didn't ice the bottom of the cake and filled the top with the ganache. The design process can be a little tricky, but it is completed with a simple and sophisticated look with a clear contrast between black and white.

Sponge Cake Recipe

by Cho Eunih

Make your cake more perfect with delicious sponge cake recipes. I will show you genoise and chiffon cake recipes by Cho Eunih instructor from the Cake Recipe Class, joys_kitchen.

01

Genoise

size

Round cake (diameter 15 cm, height 7 cm)

ingredients

Plain: 135 g whole eggs, 100 g sugar, 90 g cake flour, 25 g milk, 15 g butter, 10 g corn starch

Chocolat: 150 g whole eggs, 110 g sugar, 75 g cake flour, 25 g cocoa powder, 22 g butter, 15 g milk

Matcha: 150 g whole eggs, 100 g sugar, 90 g cake flour, 20 g milk, 15 g butter, 7 g matcha powder

point

❶ The recipes of chocolat and matcha genoise are made the same way as plain genoise.

❷ Because the genoise containing cocoa powder or matcha powder deflates easier than the plain genoise, whip until the structure becomes harder and denser before the powder ingredients are mixed in. Make sure to mix the batter quickly so that the air pockets don't deflate as much as possible.

❸ A well-baked genoise does not shrink and neatly separates from the parchment paper after cooling.

preparation

❶ Sieve all the powdered ingredients in advance.

❷ Scale the butter and milk together, melt and keep warm in a water bath.

❸ Cut and line the Teflon sheet or parchment paper to fit the cake pan.

❹ Preheat the oven at 170°C in advance.

cake recipe

1   Lightly whisk the whole eggs, and mix in the sugar.

2   Warm to 35~40˚C on a bowl filled with warm water.

3   When the temperature reaches 35~40˚C, remove from the water bath and whip the egg mixture at medium speed to aerate.

4   When it volumize as it collects air, whip at low speed to make the foam dense and smooth overall.

5   Add the sifted powders and mix with a rubber spatula.

6   Mix quickly until no dry ingredients are visible, but work carefully, taking care not to deflate the foam.

1   2   3

4   5   6

cake recipe

7   Add previously melted and warmed butter and milk, and mix quickly.

8   When the batter is mixed smoothly, pour into the cake pan lined with parchment paper.

9   Lower the preheated oven temperature from 170˚C to 160˚C, and bake for 25~30 minutes. Remove from the pan and cool on a cooling rack.

7   8   9

02

Chiffon

size

Chiffon cake (diameter 15 cm, height 10 cm)

ingredients

40 g egg yolks, 15 g sugar A, 60 g cake flour, 1/3 ts baking powder, 42 g butter, 40 g milk, 15 g condensed milk, pinch of salt, 110 g egg whites, 65 g sugar B

point

❶ Chiffon is iced all the way to the center hole, so make sure to clean the inside, so no crumbs remain.

❷ A cake with a recessed upper part is difficult to ice. Bake the cake well according to the time in order to remove neatly from the pan, and cool sufficiently before removing.

preparation

❶ Sieve all the powdered ingredients in advance.

❷ Scale the butter, milk, and condensed milk together, melt and keep warm in a water bath.

❸ Preheat the oven at 170°C in advance.

cake recipe

1   In a bowl, add egg yolks and sugar A, and whisk until the mixture turns pale yellow.

2   Add previously melted and warmed butter, milk, and condensed milk with a pinch of salt, and lightly mix.

3   Sieve cake flour and baking powder into 2.

4   Mix until no dry ingredients are visible.

5   Cover with a wet towel until needed.

6   Whip egg whites, gradually adding sugar, to make a firm and dense meringue.

1   2   3

4   5   6

7   Divide and add the meringue in three parts and mix with 5 using a rubber spatula into a smooth batter.

8   Pour into a chiffon pan and bake for 32~35 minutes in a preheated oven at 170˚C. Cool the cake in the pan turned upside down.

9   Neatly remove the cooled chiffon by scraping the inner pillar, bottom, and edges with a chiffon spatula.

7   8   9

Editor's
Pick
Cake design

There are already many books on cake design available. Flower cakes made with buttercream, and Tteok (rice cake) cakes made with white bean paste. All these are cake designs that have been popular past 2~3 years. Among many types of cakes, I started planning with the idea of the most favored cake design available in any dessert café, cakes that do not go out of trend- and thought, 'I wish there is a technical book on whipped cream cake design.'

Editor's Pick
<u>Cake design</u>

    As it is a book that contains whipped cream cake design skills, I wanted to include theoretical content and designs that reflect the latest trends. Not only so but at the same time being able to learn efficiently and applied practically by those who do baking as a hobby and even by professionals in the industry who runs a dessert café.

    Instructor Jung Hayeon is very popular in this field that she teaches not only in Korea but also overseas, such as in China, Taiwan, and Singapore. In order to include all the skills related to the handling of whipped cream and design of the cake in this book, the manuscript was revised, supplemented and photos have been retaken numerous times to make it easy to understand all the know-how a book can contain. I hope all the beautiful works that bloomed in her hands will also bloom in the hands of the readers who will utilize this book.

Inasmuch as the instructor Jung Hayeon and all those who helped with this book have endured for many months, I was able to complete a book that contains design techniques that have never existed but truly necessary.

I am sincerely grateful to the instructor Jung Hayeon for writing her know-how into the manuscript despite her busy schedules. Thanks to her devotion, this book was able to be published with nothing to waste- not a single character, a sentence, nothing. I also want to thank instructor Cho Eunih for helping us make the shooting go easier and providing delicious cake recipes. (You will join us next time as well?) Additionally, I would like to express my gratitude to the photographer Kim Namhun, who took over 5,000 pictures for a week, and the designer Chang Jiyoon who made it easy to follow more than 300 pages as if watching a video.

December 2020, THETABLE- Planning Team

Cake Design Recipe *Diary*

Cake Design Recipe *Diary*

CONGMOM'S CAKE DIARY SERIES

English Version

English Version

Korean Version

GARUHARU MASTER BOOK SERIES

ÉCLAIR
by GARUHARU

TARTE
by GARUHARU

DECORATION
by GARUHARU

* These books are written side-by-side in both Korean and English.

You can buy Congmom's books and GARUHARU books
more easily and cheaper on the K-zone studio site.
(www.kzonestudio.com)